Acknowledgments

Many talented individuals have contributed to the creation of the IWS Prep course. We would like to thank Maurizio Broggi and Kirra Barnes, who spearheaded this project and developed the framework for the program. Thank you to Andrea Eby, Italian Wine Program Director, for taking this book to the finish line. The fabulous maps found within the IWS Prep course book are thanks to Quentin Sadler, mapmaker extraordinaire. We also thank Darren Lingard, graphic artist, for his wonderful work on the winemaking and viticulture diagrams. Thank you to Kate Hall for her meticulous editing skills and attention to detail. Thank you to Lisa Airey, Education Director, for her invaluable leadership in shepherding this project to completion.

To all who have helped and supported us with this educational initiative, our heartfelt thanks.

To you, IWS Prep candidate, good luck and good reading.

Welcome to the Italian Wine Scholar Prep course, The Wines of Italy Glass by Glass!

The Wine Scholar Guild is thrilled that you have chosen to begin your Italian wine education journey with us!

This course has been designed to provide students with a fundamental understanding of the Italian wines most often seen on a retail shelf and restaurant wine list. This is not an exhaustive compilation. Instead, wine styles and regions have been carefully chosen from Northern, Central and Southern Italy to reflect their commercial and historical significance.

Information has been judiciously curated. Icons help to signpost your learning. Information on "the place," "grape varieties," "wine profiles" and "production notes" has been included for each wine. We have also included a "notable producers" section and let us just say that creating this list was no easy task! Each of the appellations has dozens, if not hundreds, of excellent producers that could merit a mention. Instead, we have chosen a cross-section of producers that represent the diversity of their particular appellation. We have included large producers (whose wines will be widely available), boutique producers, historically significant wineries and up-and-coming entities. If one of the focus wines is unavailable in your market, the "Detour" wines make the perfect substitution, or… you can use the "Detours" to extend your virtual vacation and taste even more of what a region has to offer. Peppered with information on cultural attractions, regional cuisines, and points of interest, the program allows students to learn about wine within a cultural framework.

We see IWS Prep as a springboard toward a comprehensive understanding of the wines of Italy. For those of you who have registered to take part in an online or classroom session, please be sure to take advantage of the E-learning modules that accompany this coursebook! The modules are designed to support and enhance your learning, be it classroom-based or independent study.

Upon completion, you will be ideally positioned to further your studies in the internationally acclaimed Italian Wine Scholar (IWS)™ Certification Program! (Learn more on the last page.)

We trust you will enjoy your scintillating excursion through Italy, glass by glass!

Salute!

Handwritten notes:

For each section:
1) read text
2) review materials on online module
3) review live webinars

Exam: online, min score of 60pc
50 questions - multiple choice
45 mins to complete

TABLE OF CONTENTS

Phoenicians plant
grapevines
throughout the
Mediterranean

15th - 3rd c. BCE

Greeks arrive and
name Italy 'Oenotria,'
meaning
'the land of vines'

8th - 6th c. BCE

Fall of Roman Empire;
wine industry enters
a period of profound
regression

476 CE

Rinascimento
(Italian Renaissance)
begins; the practice of
mezzadria (sharecropping)
becomes widely
adopted

13th c. CE

Christopher Columbus's
voyage to the
Americas

1492

Michelangelo
finishes painting
the ceiling of the
Sistine Chapel

1512

Il Risorgimento
begins (a period of
political and social
rebirth)

1815

7th c. BCE

Etruscans develop
extensive trade network
and export wines to
present-day France
and Spain

3rd c. BCE - 5th c. CE

Romans supersede
Greeks as the major
political and military
power of the
ancient world

6th - 13th c. CE

During the Dark Ages,
viticulture and wine
culture are promulgated
by monasteries

15th c. CE

The first literary
mentions of important
Italian grapes
(e.g. Nebbiolo)

1506

Leonardo
da Vinci
paints the
Mona Lisa

1720

Kingdom of
Sardegna
forms

INTRODUCTION TO ITALY

Italy…few countries are as closely linked to the history of wine! For well over 2,000 years, Italia has been growing grapes and making wine. When the ancient Greeks arrived on the shores of present-day Campania, they found a thriving viticultural scene and named the land "Oenotria," or "land of the vine." Italy has been blessed with a climate that vines adore. The country is a treasure trove of grape varieties… the number of native cultivars is estimated to be between 350 and 600! With hundreds of varieties in commercial production and what seems like a complicated system of wine categorization, learning about the world of Italian wine can seem overwhelming. However, with a few key pieces of information, anyone can learn to unlock the door to the wonderful world of Italian wine!

Although an Old-World country, Italy itself is younger than most people would imagine. In the centuries prior to unification, Italy was divided into a plethora of small kingdoms and city-states. Each of these political entities shaped the history, people and culture of what would become a "region" of Italy. Almost 160 years after the Risorgimento (Unification of Italy, 1861), many Italians remain fiercely loyal to their region or even their village vs. their country. Extreme neighborhood pride is referred to as campanilismo. The term is taken from the word "campanile" meaning "bell tower," i.e. every Italian supports their own church's bell tower.

Regions also have their rivalries, but there is no greater division within the country than that of north and south. Some northerners think southerners take an inordinate amount of joy out of life's little pleasures, while many southerners think northerners put too much focus on their work life vs. their personal life. Of course, central Italy lies between these polar opposites, offering a unique amalgamation of both. A journey across the country is really more like a trek across three cultures, 20 countries, and 1000s of self-proclaimed, independent city-states.

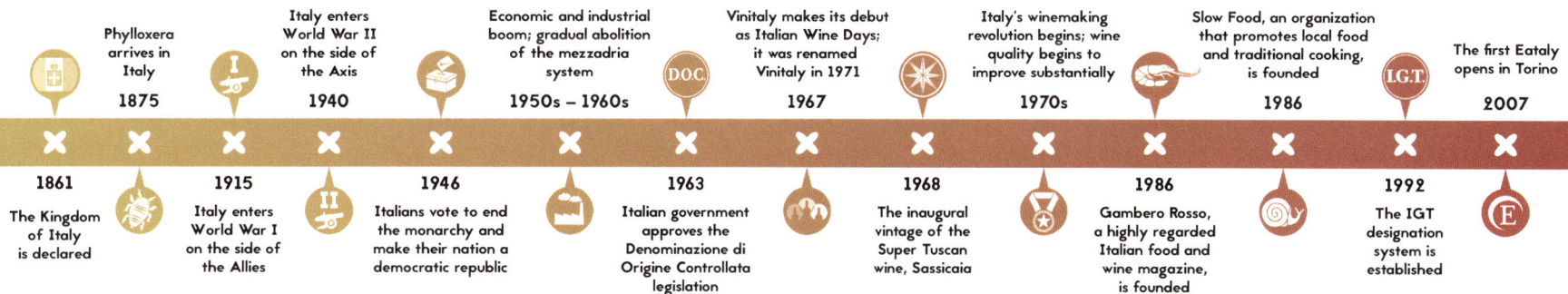

Phylloxera arrives in Italy
1875

Italy enters World War II on the side of the Axis
1940

Economic and industrial boom; gradual abolition of the mezzadria system
1950s – 1960s

D.O.C.

Vinitaly makes its debut as Italian Wine Days; it was renamed Vinitaly in 1971
1967

Italy's winemaking revolution begins; wine quality begins to improve substantially
1970s

Slow Food, an organization that promotes local food and traditional cooking, is founded
1986

I.G.T.

The first Eataly opens in Torino
2007

1861
The Kingdom of Italy is declared

1915
Italy enters World War I on the side of the Allies

1946
Italians vote to end the monarchy and make their nation a democratic republic

1963
Italian government approves the Denominazione di Origine Controllata legislation

1968
The inaugural vintage of the Super Tuscan wine, Sassicaia

1986
Gambero Rosso, a highly regarded Italian food and wine magazine, is founded

1992
The IGT designation system is established

GEOGRAPHY, TOPOGRAPHY, CLIMATE & SOILS

Italy is considered a southern European country. It is bordered by Switzerland and Austria to the north, France to the northwest and Slovenia to the northeast. The mainland boasts an impressive 4,598 miles/7,400 km of coastline and is surrounded on three sides by the Mediterranean Sea. The islands of Sardegna and Sicilia, plus a multitude of archipelagos, extend Italy's reach beyond the mainland.

Over three-quarters of Italy is covered by mountains and hills, and the hills that radiate out from the Alps and the Apennines are where the vast majority of vineyards are located. Since Roman times, vineyard plantings have been concentrated on the slopes, and with few exceptions, that is where the majority of them remain today.

Plains account for the smallest proportion of the Italian landscape and the flat and fertile Padan Plain accounts for more than two-thirds of that area. It is Italy's largest plain and is crossed by Italy's longest river, the Po.

SEAS, RIVERS AND LAKES

The Mediterranean Sea surrounding the Italian Peninsula is sub-divided into four major basins: the Adriatic, Ionian, Tyrrhenian and Ligurian. Each of these plays a significant role in moderating the climate of the land it borders. Italy is blessed with many rivers, but due to the topography, most are shorter and volumetrically smaller than rivers found in other parts of Europe. As a result, their impact on local climate tends to be less pronounced. Lakes, on the other hand, play a significant role wherever they are present. Famous lakes such as Lago di Garda, Lago di Como and Lago d'Iseo all impact the surrounding areas by moderating the cooling influence of the Alps.

CLIMATE *elevation is key*

Despite the fact that Italy stretches over 10 degrees of latitude, it is elevation that is key to determining the climate of its winegrowing areas. Growing seasons are extended with elevation, allowing grapes to maintain acidity and develop aromatic complexity. Because many of Italy's vineyards are planted at significant elevation, the country has some of the latest harvests in Europe, despite being surrounded on three sides by the Mediterranean. It is the interplay between mountains and sea that is the unifying and fundamental feature of Italy's climate.

SOILS

The soils of Italy are diverse. Within a single region there might be several soil types resulting from complex and different geological evolution. However, the soils of Italy can be broadly categorized based on their origin and formation.

Sedimentary - Most of Italy's vineyards are found on soils derived from bedrock of alluvial or marine origin, or deposited from glaciers (such as the glacial moraines of northern Piemonte, Franciacorta, Valtellina and the area around Lake Garda).

Metamorphic - The heat, pressure and chemical processes that occurred during the creation of Italy's mountain ranges resulted in the formation of many metamorphic rocks. Soils derived from foliated metamorphic rocks such as gneiss, schist and slate can be found in Italian vineyards, particularly in parts of Sardegna, Calabria, northeastern Sicilia and the Alps.

Volcanic - Italy's geological evolution has been strongly shaped by volcanic activity and a considerable number of vineyards are sited on the slopes of extinct or active volcanoes. Due to the presence of volcanic ash, the soils are often nutrient-rich and water-retentive.

The European Union (EU) regulates the wine industry in all of its member states. In 2009, new regulations designed to standardize the nomenclature of quality levels and labelling terms were introduced. The EU divided wine into two major groups and created new categories that roughly corresponded to the quality pyramids already in place in member countries:

Wines with geographical indication

- Protected Designation of Origin (PDO)/Denominazione di Origine Protetta (DOP)
- Protected Geographical Indication (PGI)/Indicazione Geografica Protetta (IGP)

Wines without geographical indication (generic wine)

Many countries, including Italy, chose to continue to use their traditional designations for each of these categories, as they generally fit neatly into the EU structure. The majority of Italian wines are labeled using the traditional designations (DOCG, DOC, and IGT), which continue to be the most common quality designations found on labels today.

DOCG – Denominazione di Origine Controllata e Garantita
- Wines at the very top of the wine quality pyramid
- Wines of high reputation showing intrinsic qualities inherent to their specific delimited area of production; these production zones are usually smaller than most DOCs.

DOC – Denominazione di Origine Controllata
- One step in quality beneath the DOCG level
- The wines come from delimited geographical area(s) that are usually larger in size than that of the DOCGs

IGT – Indicazione Geografica Tipica
- Wines are primarily defined by an indication of the geographical area where they are made
- Growing areas are generally quite large, ranging from an entire province or region to multiple regions or provinces

Vino – Generic Wine
- Wines that do not have an indication of origin other than the country in which they were made
- Generally produced from grapes grown outside appellation boundaries or a blend of grapes from multiple appellations

Classico – A wine produced from the original historic winegrowing area of a DOCG or DOC

Superiore – A wine with a higher minimum actual alcohol content compared to the non-superiore version of the same wine; often superiore wines also have stricter production criteria than non-superiore wines

Riserva – A wine that went through an extended period of aging before release compared to the non-riserva version; a riserva wine usually has stricter production requirements as well

Rosso – A red wine

Bianco – A white wine

Rosato – A rosé wine

Spumante – A fully sparkling wine

Frizzante – A semi-sparkling wine

Passito – A wine made from semi-dried grapes

Secco – A dry wine (max 0.4% or 4g/l of sugar)

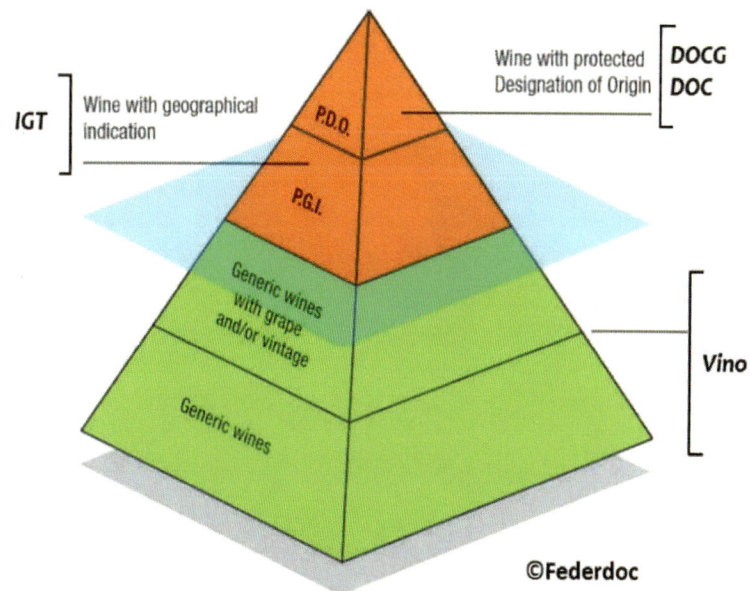

©Federdoc

At first glance, many Italian wine labels appear daunting. The names of grape varieties, wine regions, stylistic terms and fantasy names can all be included on a label; determining the meaning of all this information can be challenging. There are three different ways in which information is organized in order to let you know what is in the bottle:

- By place/region e.g. Barolo DOCG
- By grape + place e.g. Barbera d'Alba DOC (Note that in Italian, d' or di means "of" or "from" – i.e., Barbera from Alba).
- By fantasy name + place/region e.g. Montevertine Le Pergole Torte Toscana IGT

1 – Appellation (name of the winegrowing zone)
2 – Quality designation e.g. DOC, DOCG (DOP) or IGT (IGP)
3 – Country of Origin
4 – Vintage
5 – Name and location of bottler
6 – Indication of the batch
7 – Alcoholic strength
8 – Volume of wine

No other country has quite the abundance of native grape varieties as does Italy. The historical isolation of many of Italy's winegrowing areas allowed native grapes to survive into the modern age. The plethora of indigenous grape varieties is a major advantage for Italian producers because they are able to offer unique and distinctive wines that cannot be replicated elsewhere. Descriptions have been provided for a select few of the most commercially and culturally significant varieties.

RED VARIETIES

Aglianico is considered one of Italy's noble grapes. The wines are dark, powerful, full-bodied and high in alcohol and tannin, yet this muscular profile is lifted by notably high acidity.

Barbera is Piemonte's most widely planted variety. The grape is hallmarked by high acidity. deep color, bright red cherry fruit, and low to moderate tannin.

Cannonau is genetically identical to Spain's Garnacha. It produces full-bodied wines with high levels of alcohol and both floral and red fruit aromas.

Corvina is one of the primary grapes of the Valpolicella region. Wines are moderate in tannin with aromas of sour cherry, violet, herb and almond.

Dolcetto makes deeply-colored, moderately low-acid wines with fragrant black fruit aromas. Its tannins are ample but soft and round; they deliver a pleasantly bitter finish.

Nebbiolo is one of Italy's most noble red grape varieties. In youth, the wines show aromas of red cherry, rose, violet, licorice and underbrush; with age, nuances of dried red fruit, rose petals, sweet spice, leather and truffles develop. The wines possess high levels of acid, tannin, alcohol and extract.

Nerello Mascalese is native to Etna. The wines are pale to moderately saturated in color with aromas of red fruit, herb and spice. Tannins are perceptible but smooth; the alcohol is high and the acidity is lively. It is often compared to Nebbiolo and Pinot Noir.

Nero d'Avola is Sicilia's predominant red grape. Wines can have deep intensity of color, fine smooth tannins, soft texture, full body and high alcohol—all balanced by fresh acidity. Aromas and flavors include cherry, plum, blackberry and Mediterranean brush and herbs.

Montepulciano has thick skins which produce wines that are deep ruby in color. The wines tend to be robust, full-bodied and high in alcohol with dense, ripe tannins and overt red cherry and plum fruit.

Primitivo is rich in anthocyanins. It produces big, alcoholic wines laced with aromas of cherry, raspberry, tobacco and herbs.

Sagrantino is Umbria's most distinctive red grape. It is notably tannic, deep in color, full-bodied and powerful with lively acidity and high levels of extract and alcohol.

Sangiovese is Italy's most widely planted grape variety. It produces wines that tend to be lighter in color with high acidity, a noticeable tannic grip and violet, sour cherry, plum and tealeaf aromas.

WHITE VARIETIES

Arneis produces medium- to full-bodied wines that boast subtle aromas and flavours of white flowers, stone fruit and ripe pear.

Cortese is one of Piemonte's principal white grapes. The wines are delicately aromatic with refreshing acidity, minerality and a fresh lemon zest character.

Fiano is listed among Italy's finest white varieties. The wines display aromas of acacia, citrus, apple, herb, hazelnut, balsam, honey and mineral. The wines age particularly well, developing complexity and intense flinty, smoky notes.

Garganega is one of the most ancient grape varieties in Italy. The wines possess a steely acidity marked by aromas of white flowers, citrus, ripe yellow fruit, almonds and minerals.

Glera is the principal grape of Prosecco. It produces wines that are light and refreshing with aromas of white flowers, lemon, pear, apple and peach.

Grechetto di Orvieto is considered the traditional Grechetto grape of Umbria (there is a another!). This Grechetto produces light citrusy wines with refreshing acidity.

Greco stands among the most ancient and finest white grapes of Campania. Greco wines are well structured with marked acidity and a round, full-bodied texture. Often, they demonstrate an almost tannic mouthfeel.

Moscato Bianco also known as Muscat Blanc à Petits Grains, is the most widely planted white grape in Piemonte. The wines show the full aromatic profile of Moscato with pronounced floral and fruity aromas (orange blossom, stone fruit, citrus) combined with notes of honey, musk and spice.

Verdicchio is one of Italy's noble varieties ; the wines have attractive floral and citrusy aromas with a mineral core and almond finish. They are well-structured with bracing acidity and high levels of extract and alcohol.

Vernaccia di San Gimignano is an ancient Tuscan variety grown around the town of San Gimignano. It produces zesty, mineral wines with a nut-skin finish.

Vermentino wines display intense floral and fruity aromas along with aromatic herbs and stone/mineral notes. Often high in alcohol, the wines are balanced by a refreshing acidity and a saline finish.

INTERNATIONAL GRAPE VARIETIES

Despite Italy's wealth of native grapes, international varieties have also played an important role in the country's viticultural development. French and German varieties were introduced to Italy as early as the 19th century. A second influx of international varieties took place after the outbreak of phylloxera (a louse that decimated European vineyards). In the waves of replanting that ensued, many indigenous varieties were replaced with French varieties and

consequently lost forever. More international vines were planted in the 1970s and 80s. Toscana became a major center for international grape varieties by the 1990s. In recent years, however, the fashion for international varieties has fallen considerably. Producers are now more interested in making wines in which native grapes can shine.

WINE STYLES

Italy competes with France as the world's largest wine producer. It has one of the largest areas under vine, behind Spain and France. Vines are grown in all of Italy's 20 administrative regions. With hundreds of grape varieties in the ground, wine styles are diverse.

- Red and white wines are both well represented with slightly more white wine production than red. Rosés remain a fairly minor category; however, several areas are well recognized for their production.

- Over the past 30 years, Italy has significantly increased the quality and quantity of the sparkling wine it produces; it is now the world's largest sparkling wine producer

- Italy has a long tradition of sweet wine production and examples can be found throughout the country

ITALIAN WINE PRODUCTION BY CATEGORY (2018)

■ PDO ■ PGI ■ Vino

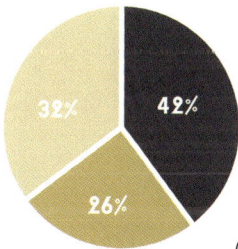

42%
26%
32%

ITALIAN WINE PRODUCTION BY COLOR (2018)

■ Red/Rosé ■ White

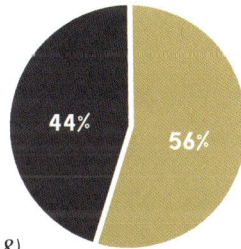

44%
56%

(Data Source: ISTAT 2018)

THE CUISINE OF ITALY

The cuisine of Italy has been shaped by its history. Ancient Rome developed the Mediterranean diet of wine, olive oil and bread. Exotic items such as butter and beer were introduced by the invading Barbarian hordes; pasta, spices, sugar and dried fruit were introduced by the Arabs. With the 'discovery' of America in 1492, novelties from the New World began to be incorporated into Italian cuisine. Turkey, sweet corn, potatoes, and of course, the tomato, were all adopted. During the 17th century, the idea of "Italian" food was somewhat solidified as chefs tried to create a national cuisine. By the 1800s, regional cuisines were being celebrated.

Pizza, pesto and pasta alla carbonara were conceived during the 19th century and remain some of the country's most iconic foods. It was also at this time that Pellegrino Artusi published his Manuale dell'Artusi: La Scienza in Cucina e l'Arte di Mangiar Bene (The Artusi Manual: Science in the Kitchen and the Art of Eating Well). This landmark publication included traditional recipes from different regions of the country. For the first time, the dishes of the average Italian were documented, not just those of Italy's rich and famous.

Two world wars, the emergence of women in the workplace and industrialization have all had a significant impact on Italy's cuisine—fast food restaurants and processed foods made their appearance. But while these options do have a presence in Italy, many Italians have worked hard to counteract their influence. Cultural movements, such as Slow Food, aim to preserve Italy's culinary traditions, but more importantly, Italy's fierce pride in its culinary traditions tucks regional treasures under the umbrella of a national cuisine and keeps them well protected.

AUSTRIA

Brenner Pass

TRENTINO-ALTO ADIGE

Adige

Bolzano /Bozen

DOLOMITES

JULIAN ALPS

SLOVENIA

FRIULI VENEZIA GIULIA

Udine

Gorizia

Pordenone

SWITZERLAND

Monte Cervino (Matterhorn) 14,692 ft/ 4,478 m

THE ALPS

Lago di Como

THE ALPS

Trento

VENETO

Monte Bianco (Mont Blanc) 15,774 ft/ 4,808 m

VALLE D'AOSTA

Aosta

Lago Maggiore

Lago d'Iseo

Bergamo

Lago di Garda

Vicenza

Treviso

Trieste

ADRIATIC SEA

Brescia

Verona

Padova

Venezia

PIEMONTE

Milano

LOMBARDIA

Po

Pavia

Cremona

Mantova

Adige

Torino

Po

Asti

Alessandria

Parma

EMILIA-ROMAGNA

Po

Ferrara

Comacchio

Alba

Modena

Po

F R A N C E

Cuneo

Genova

Bologna

Ravenna

A P E N N I N E S

MARITIME ALPS

Portofino

LIGURIA

GULF OF GENOA

LIGURIAN SEA

La Spezia

Rimini

T O S C A N A

MARCHE

Map by Quentin Sadler

WINE SCHOLAR GUILD

0   50   100 km

0   25   50 miles

N

NORTHERN ITALY

Northern Italy encompasses eight administrative regions located in the northernmost part of the country. It has a history of Germanic, French and Gallic rule and is often collectively referred to as "Germany" by southern Italians. Like its border countries, northern Italy is continental in climate—as opposed to the remainder of the peninsula, which is Mediterranean.

But weather is not the only distinction between the north and the rest of the country. Northern Italy accounts for almost half of Italy's total population, produces more than half of its national gross domestic product (GDP) and has one of the highest GDPs per capita in Europe. In fact, northern Italy is the economic, financial and industrial engine of Italy.

Northern Italy is also home to some of Italy's most famous destinations and attractions. The city of Venezia, with its legendary lagoon, is one of the most visited cities in Italy and is universally considered among the world's most beautiful. Truman Capote once remarked "Venice is like eating an entire box of chocolate liqueurs in one go." Those lucky enough to visit often feel similarly overindulged by the sheer beauty of this floating city.

Verona is also a very popular tourist destination with its highly polished monument dedicated to Romeo and Juliet and its beautifully preserved Roman amphitheater, which still hosts spectacular concerts and opera performances. Milano is considered the most cosmopolitan of Italy's cities. Luxury fashion houses such as Armani, Prada, Versace and Dolce & Gabbana are all headquartered here and the city's Via Monte Napoleone is considered one of the world's most expensive shopping streets.

The cuisine of northern Italy is deliciously distinct from the remainder of the peninsula. Perhaps the most obvious difference is the predominance of butter rather than olive oil. Tomatoes also take somewhat of a back seat to stocks, cream and wine in sauces. One is also more likely to find stuffed pastas (such as ravioli) versus the flat and extruded pastas common further south; moreover, polenta and risotto (often accompanied by the region's famous white truffles) are often the star starch of the meal.

The north is also known for the quality of its beef and pork products. Meats are cured into savory salumi or stewed, roasted and braised. Dairy products are also highly prized. In fact, some of Italy's most famous cheeses including Gorgonzola, Parmigiano Reggiano and Grana Padano all hail from the north. Seafood and freshwater fish are also common additions to the table. The eels of Comacchio are famous, as are the mussels and clams of Veneto's coast.

Many dishes in the north are rich and hearty. There is food and wine... and work. But here, these three things make a happy triumvirate.

N. Italian Sparkling - Prosecco DOC
large territory including all of
Friuli-Venezia Giulia & most of
Veneto.
DOCG smaller area N. of Treviso. foothills of
Alps. slopes trap sun and warmer winds.

NORTHERN ITALY
Sparkling Wines

NORTHERN ITALY SPARKLING WINES

50 100 km

0 25 50 miles

AUSTRIA

SWITZERLAND

FRANCE

SLOVENIA

TRENTINO-ALTO ADIGE

FRIULI VENEZIA GIULIA

Adige

Bolzano /Bozen

Trento DOC

Trento

Prosecco Superiore DOCG

Lago di Como

Lago Maggiore

VALLE D'AOSTA

Valle d'Aosta

Aosta

LOMBARDIA

Lago d'Iseo

Lago di Garda

VENETO

Treviso

Prosecco DOC

Trieste

Franciacorta DOCG

Brescia

Baccalà

Spritz

Vicenza

Milano

Verona

Padova

Venezia

PIEMONTE

Vermouth

Panettone

ADRIATIC SEA

Po

Po

Panna Cotta

Prosciutto di Parma

Parmigiano Reggiano

Lambrusco Salamino di Santa Croce DOC

Adige

Torino

Moscato d'Asti & Asti DOCG

Asti

Parma

Ferrara

Po

Alba

Reggiano Lambrusco DOC

Lambrusco di Sorbara DOC

Modena

Aceto Balsamico Tradizionale

Lambrusco di Modena DOC

Bologna

EMILIA-ROMAGNA

Ravenna

Genova

Portofino

Lambrusco Grasparossa di Castelvetro DOC

Tortellini

Italian Riviera

Cinque Terre

GULF OF GENOA

LIGURIAN SEA

LIGURIA

TOSCANA

Lasagne

Rimini

MARCHE

Map by Quentin Sadler

WINE SCHOLAR GUILD

18

NORTHERN ITALY SPARKLING WINES:
THE ROAD MAP

1 PROSECCO DOC & PROSECCO SUPERIORE DOCG

Proh-SEHK-oh;
Proh-SEHK-oh Soo-pehr-EE-oreh

The region: mostly VENETO
VEH-neh-toe

The region of Veneto is one third of the "Tre Venezie" or "Triveneto" (three Venices), a historical area encompassing Veneto, Trentino-Alto Adige and Friuli Venezia Giulia. The Alps define Veneto's north, but hills give way to a plain leading to the Adriatic Sea.

Not to miss travel site:

Venice is Veneto's famous and historically important capital city. As a key medieval port city, its traders introduced much of Europe to Middle and Far Eastern spices, perfumes and textiles.

4 ASTI DOCG & MOSCATO D'ASTI DOCG
AHS-tee; Mos-KAH-toh DAHS-tee

The region: PIEMONTE/PIEDMONT
Pee-ay-MON-tay

The name Piemonte means "at the foot of the mountain." The region is aptly named, as it is surrounded on three sides by mountain ranges. The mountains fade into impressive hills which in turn give way to a plain in the middle of the region.

Detour to
TRENTO DOC
Trehn-toe

The region: TRENTINO-ALTO ADIGE
Trehn-TEE-noh AHL-toh AH-dee-jeh

Trentino is administratively tied to its neighbor Alto Adige as the region of Trentino-Alto Adige, but each maintains an independent identity. Trentino is very much Italian in mindset, while Alto Adige is aligned with German culture and language. However, both these sub-regions are, perhaps first and foremost, Alpine wine regions - both provinces are dominated by the Italian Alps.

Not to miss travel site:

Canelli is the heart of Asti production and its "Underground Cathedrals" refer to the extensive network of tunnels and caves underneath the city. Carved during the 16th to 19th centuries, they were adopted as wine cellars. Stlll largely in use, there are four historic cellars that can be visited by the public.

---

2 FRANCIACORTA DOCG
FRAN-chee-ah-cor-ta

The region: LOMBARDIA/LOMBARDY
Lom-bahr-DEE-ah

Lombardia shares the Alps with Switzerland. The Alpine range gives way to a narrow band of lower-elevation mountains and hills as the terrain gently yields to the fertile Padan Plain. Lombardia is one of Italy's most populated regions and the wealthiest and most industrialized in the country.

Not to miss travel site:

Milan is not only the region's capital, but is also Italy's capital of fashion, finance and industry.

3 LAMBRUSCO DOCs
Lahm-BROOS-koh

The region: EMILIA-ROMAGNA
Eh-MEE-lee-ah Roh-MAH-nyah

The Po River separates Emilia-Romagna from Lombardia and Veneto. Nearly half of Emilia-Romagna is covered by the Po River Valley, a flat, fertile plain where sixty percent of the region's wine is made. Emilia-Romagna is actually two formerly independent regions joined together administratively. However, each region retains its unique identity with distinct traditions and cultures.

Not to miss travel site:

The ancient city of Modena is one of the region's prettiest. Perhaps its most famous feature is the Piazza Grande, the central square. The centerpiece of the square is the Cathedral of Modena which houses the Ghirlandina bell tower, a masterpiece of the Romanesque style and the symbol of the city.

PROSECCO DOC & PROSECCO SUPERIORE DOCG
Veneto & Friuli Venezia Giulia

In just a few years, the Prosecco DOC has become the world's biggest sparkling wine appellation (by volume), not to mention the most recognized and popular fizz. The success of Prosecco DOC has somehow overshadowed the higher-quality sparkling wines of Prosecco Superiore DOCG made in the smaller and unique area of Conegliano Valdobbiadene (*KOH-neh-LEEYAH-noh VAHL-dohb-BEE-ah-di-nay*). This latter category really merits attention.

THE PLACE

Prosecco DOC is made in a large swath of territory covering most of the Veneto region and the entire region of Friuli Venezia Giulia. Most of the production, however, takes place in the Treviso province in Veneto. The vineyards of the Prosecco DOC are largely planted on flat plains and generate a tremendous amount of easy-to-drink sparkling wines. In fact, it is Italy's largest DOP by volume (by more than 400 million bottles!).

The Prosecco Superiore DOCG (formally designated as **Conegliano Valdobbiadene Prosecco Superiore DOCG**) is made in a much smaller area to the north of Treviso at the foothills of the Alps. The DOCG lies in the steep hills between the towns of Conegliano and Valdobbiadene. This is the historic and original heart of Prosecco production and encompasses the best vineyard sites and growing conditions in the area.

The Alps shield the vineyards from exposure to cool north winds, while the hills help to trap some of the warm air pushed inland by the Adriatic Sea, 25 miles away. The steep, south-facing slopes play their part by holding the vineyards at just the right angle to maximize sunlight exposure – important for getting grapes ripe in cooler areas. These are high-quality wines.

GRAPE VARIETIES

Prosecco is primarily made from the white grape, Glera (*GLEH-rah*). Although the exact origin of the grape is unknown, it is believed to be native to northeast Italy.

The success of Prosecco has made this grape one of Italy's most planted varieties, but virtually all of these plantings are found in Veneto and Friuli.

THE PROSECCO NAME

Before 2009, Prosecco was the name of the grape used to produce Prosecco sparkling wine. In 2009, Prosecco was adopted as the name of the appellation (Prosecco DOC) where the grape is cultivated. Glera, a local Friulian synonym, was chosen as the new, official name of the grape. The change allowed Prosecco to become a designation of origin so that producers from other regions could not produce and market Prosecco wines. Not everyone complied. Australia continues to make varietally labeled Prosecco.

CARTIZZE

Cartizze is the only sub-zone of the Conegliano Valdobbiadene Prosecco Superiore DOCG. Considered to have the finest and steepest vineyards in the appellation, Cartizze produces some of the appellation's best and most distinctive wines. Wines from the sub-zone are usually identified as "Valdobbiadene Superiore di Cartizze DOCG."

RIVE

The word "rive" translates as "vineyards on steep slopes." Forty-three villages are allowed to append their village name to the word "rive" on wine labels (provided the wines meet required production guidelines). Rive sites are second only to Cartizze in reputation and offer some of Conegliano Valdobbiadene's highest-quality wines.

WINE PROFILE

What makes this wine special: Prosecco offers alluring floral aromas, tantalizing fruit flavors and a soft, creamy mousse; this wine is a true crowd-pleaser

Acidity: Moderate

Body: Light- to medium-bodied, moderate in alcohol

Common descriptors: White flowers, citrus, peach, pear, apple, soft, fruity

Food pairings: Aperitif, appetizers; it is an excellent sparkling wine for cocktails

NOTABLE PRODUCERS

Carpenè Malvolti: The founder, Antonio Carpenè, pioneered production of sparkling Prosecco in the 19th century

Mionetto: One of the region's foremost Prosecco producers; also one of the area's oldest (founded in 1887)!

La Marca: La Marca was founded as a cooperative in 1968; today, its winegrowers cultivate more than 22,239 ac/9,000 ha of vines and it has become a top-selling brand

Ruggeri: Ruggeri is considered one of the benchmark producers and is highly respected; it has been focused on quality since its inception

PRODUCTION NOTES

Prosecco sparkling wines are usually made via the tank method (also known as the Martinotti method). This method is considered ideal to enhance the floral and fruity character of the Glera grape without adding the intense yeasty/toasty character associated with lees aging. The wine does not rest upon the lees but is filtered and bottled (under pressure) shortly after the second fermentation has been completed.

Prosecco DOC

- Most Prosecco DOC wines are made with 100% Glera, but up to 15% of other permitted grapes are allowed
- The majority of Prosecco DOC wines are fully sparkling
- Extra-dry (i.e. with some residual sugar) is the most popular style within the Prosecco DOC, with brut (drier) a distant second

Prosecco DOC Rosé

- This newly approved category will allow for the addition of 10%-15% Pinot Nero to produce sparkling Prosecco Rosé
- The wines can be produced with sweetness levels brut nature through to extra dry

Conegliano Valdobbiadene Prosecco Superiore DOCG

- Most Prosecco Superiore DOCG is made with 100% Glera, but up to 15% of native blending grapes are allowed
- The overwhelming majority of the wines in the DOCG are fully sparkling (spumante) and only the fully sparkling style carries the 'Superiore' designation in the name
- Extra-dry is the most popular style, but brut is also common and well regarded

VENETO'S SIGNATURE CUISINE

FISH

Veneto has a varied and diverse food scene - a glorious combination of mountain, plain and coastal foods. Venice and surrounding areas (especially Chioggia) have renowned fish markets.

Sarde in saor: The term "saor" translates to "flavor" or "taste" in Venetian dialect and refers to an ancient technique of frying foods and then preserving them in a marinade of cooked onions, vinegar, sugar, pine nuts and raisins. The method was originally developed by fishermen looking to preserve their catch before the advent of refrigeration. Sardines (sarde) are the traditional base of this dish, but many other ingredients can be prepared "in saor." Sarde in saor are a popular antipasto in the bacari (taverns) of Venice.

Granseola alla Veneziana is a dish made with large spider crab fished from the Venice Lagoon. The crab is boiled and seasoned with olive oil, garlic, parsley and lemon.

DETOUR

TRENTO DOC
Trentino-Alto Adige

Trento DOC is an appellation dedicated to the production of high-quality traditional method sparkling wines which are considered among Italy's finest. They are unique because the grapes are grown at high elevations in the Alpine region of Trentino. In fact, these wines are nicknamed "bollicine di montagna" (mountain bubbles).

TRENTODOC
Sparkling Wines from the Mountains

THE PLACE

Trento DOC is made in the Trentino wine region of Italy's northeast sector and is named after Trento, the capital city of Trentino. Vineyards are planted on the slopes of the Alps up to 2,625 ft/800 m in elevation. At night, cool mountain air descends over the vineyards preserving the grapes' natural acidity.

During the day, warm air rises from the lower valleys ensuring the grapes continue to ripen. The Trentino vineyard landscape is shaped by the pergola, the traditional overhead vine training system. The vine canopy is trained horizontally, high off the ground. Grapes are picked by workers standing beneath.

TRENTINO-ALTO ADIGE

Adige

Bolzano/Bozen

Trento

Trento DOC

Prosecco Superiore DOCG

V E N E T O

Lago di

Treviso

GRAPE VARIETIES

The wines are primarily crafted from Chardonnay. Pinot Nero (Pinot Noir), Pinot Bianco (Pinot Blanc) and Meunier are also permitted.

TRADITIONAL METHOD (METODO CLASSICO)

In the making of traditional method sparkling wines, grapes are hand-picked and fermented into a still wine. The wine is then blended with other still wines to create a cuvée. This base wine is then bottled with a small amount of yeast and sugar to initiate a second fermentation in a closed environment. This step produces and captures the bubbles and takes place in the same bottle from which it is later served.

WINE PROFILE

What makes this wine special: High-altitude vineyards produce high-acid grapes and high-acid wine. This, coupled with the aromatic complexity derived from extended lees aging, crafts nervy wines with a supple mouthfeel and hints of toast and biscuit.

What you need to know: Non-vintage Trento traditional method sparkling wines must spend at least 15 months on lees. Vintage wines rest on lees for 24 months and Riserva wines, 36 months. Most Trento DOC wines are white and made in a brut style, but rosé is also produced.

Trentodoc
Although the official name of the appellation is Trento DOC, the consorzio (a local trade consortium) and producers have collectively adopted the name 'TRENTODOC' (one word), which is the designation that is shown on the labels

NOTABLE PRODUCERS

Ferrari: Giulio Ferrari, founder of the famous sparkling house that bears his name, was the first to explore Trentino's potential for sparkling wine production

Cavit, Mezzacorona, Letrari, Maso Martis and **Cesarini Sforza** all produce high-quality examples of Trento DOC

TRENTO'S SIGNATURE CUISINE

MEAT

The region of Trentino offers a variety of food-themed, self-guided routes for visitors to the area. **The Strada del Vino e dei Sapori del Trentino** (The Trentino Wine and Flavors Route) guides visitors from Lake Garda to the streets of Trento as they sample local delicacies such as fresh-water trout, local **lucanica** and **ciuìga** sausages and **zambana** asparagus.

CHEESE

Those who fancy **formaggio** (cheese) would enjoy the **Strada dei Formaggi delle Dolomiti** (The Dolomites Cheese Route). The region is famous for its raw milk cheeses such as **Trentingrana DOP, Stravecchio di Fiemme and Puzzone di Moena DOP** (to name just a few)! All are made from the milk of animals that graze the area's pristine Alpine meadows.

APPLES

Finally, the Val di Non and Val di Sole (Apples and Flavors) Route takes visitors through a pastoral landscape to sample the area's fresh produce. In addition to the region's famous apples, one can sample pears, cherries, honey and the local smoked **mortandèla** sausage.

FRANCIACORTA DOCG
Lombardia

Franciacorta is arguably Italy's most prestigious traditional method sparkling wine. It has been said that it is Italy's answer to France's champagne, but truthfully, these wines are distinctive enough to stand on their own.

THE PLACE

Franciacorta is made in the narrow band of gentle hills where the Alps meet the Padan Plain. This section of Lombardia is characterized by a number of large lakes and rivers that were carved by glaciers slowly moving across the land during the last ice age. In fact, Lombardia has the largest number of lakes in the entire country!

The Franciacorta appellation lies just to the south of Lake Iseo, near the city of Brescia. When the glacier carved out this body of water, it also formed a natural "amphitheater" or shallow "bowl" to the south of the lake. This is the Franciacorta winegrowing area.

The lake has a temperature-moderating effect, mitigating summer highs and winter lows. For this reason, lakes (and rivers too!) often serve to protect the vines from dangerous frosts.

STRADA DEL VINO Franciacorta

GRAPE VARIETIES

Primarily based on Chardonnay. Pinot Nero (Pinot Noir), Pinot Bianco (Pinot Blanc) and the native Erbamat are also permitted.

ETYMOLOGY

The name Franciacorta is believed to be derived from the Latin terms "curtes" and "francae" referencing monastic lands and properties that were exempt from taxes during the Middle Ages. Francae, in Latin, means free.

FURTHER EXPLORATION

The Curtefranca DOC covers the same area as the Franciacorta DOCG, but is for still white or red wines.

Sounds good!

SATÈN

Satèn is a specific style of Franciacorta. It is a Chardonnay-based blanc de blancs (white wine made from white grapes) bottled with less pressure than the other wines of the appellation. This results in a creamier, silkier mousse.

NON DOSATO...

Dosaggio Zero, Dosage Zéro, Pas Dosé, and Brut Nature ... Any of these terms may appear on Franciacorta bottles, but they all mean the same thing—sugar is not added to the bottle after the second fermentation.

What makes this wine special: The growing area traps enough heat to fully ripen the grapes. Extended aging adds complexity and depth. Because of the warmer climate, the wines of Franciacorta are usually lower in acidity and have more pronounced fruit flavors than champagne.

Acidity: Medium to pronounced

Body: Medium-bodied, medium alcohol

Common descriptors: Lemon, green apple, honeydew melon, yeasty, toasty

Food pairings: Appetizers, seafood, pasta with cream sauce, rich and fatty dishes

PRODUCTION NOTES

Franciacorta's aging requirements are more demanding than those set for champagne.

- Non-vintage Franciacorta must spend at least 18 months on the lees (dead yeast cells)
- Vintage wines must spend at least 30 months on the lees
- Wines labeled as Riserva must spend a minimum of 60 months on the lees
- Franciacorta produces both white and rosé sparkling wines
- Most of the wines are produced in brut style; however, extra brut and brut nature are also common

NOTABLE PRODUCERS

Barone Pizzini: The first Franciacorta estate to farm organically

Bellavista: One of the historic and largest producers of Franciacorta

Guido Berlucchi: The historic producer credited with creating the modern sparkling wines of Franciacorta in the early 1960s. Berlucchi is also the largest producer by volume.

Ca' del Bosco: The producer of Cuvée Annamaria Clementi, considered by many to be one of Italy's top sparkling wines

LOMBARDIA'S SIGNATURE CUISINE

PASTA

Casoncelli is a stuffed pasta (usually with beef and parmesan) similar to ravioli and typical of Bergamo and Brescia.

The city of Mantua is known for **tortelli di zucca**, an egg pasta stuffed with a local variety of pumpkin called "zucca mantovana."

CHEESE

Grana Padano DOP is an aged cow's milk cheese made in several northern regions. The word "grana" translates to "grainy," a nod to the flaky, coarse texture of the cheese.

Gorgonzola DOP is a soft, blue, cow's milk cheese made in Lombardy and Piemonte.

Taleggio DOP is a smear-ripened cow's milk cheese made in Lombardia (in addition to Piemonte and Veneto). It takes its name from the Taleggio Valley, near Bergamo, where this cheese was originally produced.

LAMBRUSCO DOCs
Emilia-Romagna

Lambrusco is one of the most idiosyncratic and joyful wines of Italy. Genuine, drier-styled Lambrusco has tantalized Italian palates for generations and deserves a place at the table thanks to its incredible versatility with different foods.

THE PLACE

Lambrusco grapes are cultivated in the central part of the Emilia-Romagna region, more precisely in the Emilia section of the region, between the cities of Reggio Emilia and Modena.

Most Lambrusco vineyards lie on the flat Padan Plain that extends south of the Po River. Some vineyards also lie on the hilly terrain near the foothills of the northern Apennines to the south of Modena.

Vineyards on the plains tend to produce light, delicate Lambruscos, whereas hillside vineyards produce wines with more body and structure (by comparison).

GRAPE VARIETIES

There are actually quite a few Lambrusco grape varieties! Most Lambruscos are closely related to each other but are counted as distinct grapes.

The three principal Lambrusco varieties are Lambrusco di Sorbara Lambrusco Grasparossa and Lambrusco Salamino (the most widely planted). Lambrusco wines are typically made with one or more members of the Lambrusco grape family.

What makes this wine special: Lambrusco is the quintessential cheerful Italian fizzy red wine

Acidity: Pronounced

Tannin: Low to moderate

Body: Light- to medium-bodied, moderate in alcohol

Common descriptors: Floral (violet), pink grapefruit, raspberries, strawberries, cherries

Food pairings: Lambrusco, with its bright acidity, red fruit flavors and lively, creamy fizz, is the ultimate food wine—it cuts right through fatty, rich foods and purportedly aids digestion. Despite its versatility, Lambrusco perhaps shines brightest with salami and other cured meats.

NOTABLE PRODUCERS

Cleto Chiarli: The first company in Emilia-Romagna to produce and sell Lambrusco wine; pioneered the use of the tank method for Lambrusco production

Paltrinieri: Boutique producer known for their wide range of artisanal Lambruscos

Lini: Fourth-generation, family-owned producer of the first-ever Lambrusco to be included in the Wine Spectator's Top 100 Wines of Italy

Riunite: Historic cooperative producer that brought a sweeter version of Lambrusco to America, where it has remained one of the top-selling Italian red wines for more than 40 years

PRODUCTION NOTES

Lambrusco wines are commonly made using the tank method of sparkling wine production and are mostly frizzante (semi-sparkling) in style.

The principal Lambrusco DOC appellations are:

- **Lambrusco di Sorbara DOC:** These are fragrant, light-bodied wines, very pale in color (almost like a rosé). They are based on the Lambrusco di Sorbara variety grown around the hamlet of Sorbara.
- **Lambrusco Grasparossa di Castelvetro DOC:** These are the most structured of the Lambrusco wines and are based on the Lambrusco Grasparossa variety grown around the village of Castelvetro
- **Lambrusco Salamino di Santa Croce DOC:** These wines are considered as "in between" Sorbara and Grasparossa in terms of style and are based on the Lambrusco Salamino variety grown around the hamlet of Santa Croce
- **Reggiano Lambrusco DOC:** This is the Lambrusco most commonly exported and is made from one or more of the Lambrusco varieties grown in a large area around Reggio Emilia
- **Lambrusco di Modena DOC:** This wine is made from one or more of the Lambrusco varieties grown in a large area around Modena

EMILIA-ROMAGNA'S SIGNATURE CUISINE

CHEESE

Emilia-Romagna boasts one of the richest, most tasty and well-known culinary heritages in all of Italy. The regional food culture, led by cities like Bologna, Modena, Parma and Reggio Emilia (among others), is considered one of the best in the country by Italians themselves!

Emilia is proudly home to two of Italy's most renowned food specialties: **Parmigiano Reggiano cheese** and **Prosciutto di Parma ham**. Their savory and salty richness pairs superbly with the refreshing, fruity character of local Lambrusco wines.

ASTI DOCG & MOSCATO D'ASTI DOCG
Piemonte

The sweet, sparkling wines of Asti and the lightly frothy, grapey Moscato d'Asti are among the most popular Italian sparklers around the world. These wines are arguably the best sparkling expressions of the Moscato grape, and the appellation is quantitatively one of Italy's largest.

CORK

Wines made under the Asti DOCG are bottled with a significant amount of sparkle, so they need a cork that can keep this pressure well contained. The bottle is sealed with a mushroom-shaped cork like those found in bottles of Franciacorta or champagne. Moscato d'Asti has less effervescence (lower pressure) and is bottled with a straight cork like those used to seal still wine.

FURTHER EXPLORATION

Brachetto D'Acqui DOCG (Brah-KEHT-toh DAH-kwee) is a sweet, red sparkling wine made from the aromatic Brachetto grape in nearby Acqui Terme. These wines are prized for generous aromatics and bright, ripe, red-berry fruit. It is considered the red equivalent of Asti DOCG

THE PLACE

The Asti and Moscato d'Asti DOCG winegrowing area lies within the long, rolling hills of Piemonte's historic region of Monferrato. Most of the vineyards are in the southwest quadrant (called Monferrato Astigiano) on surprisingly steep grades which necessitates a manual grape harvest.

Moscato Bianco has been grown in this area of Piemonte since the Middle Ages. The very best plots are reserved for grapes destined to become Moscato d'Asti.

The vineyards of the appellation are south and southeast of the city of Asti, which gives its name to the appellation; however, the heart of the Asti production area surrounds the town of Canelli. In fact, one of the most common synonyms of Moscato Bianco is Moscato di Canelli.

GRAPE VARIETIES

Although the official name of the grape is French (Muscat à Petits Grains Blanc), the grape itself is said to have originated in Italy or Greece. In Italy, it goes by the name "Moscato Bianco." No other wine region in the world grows as much Moscato Bianco as does Piemonte.

WINE PROFILE

What makes this wine special: The wines are the quintessential expression of Moscato—capturing its intense fragrance to create an intensely floral and fruity, sweet sparkling wine

Acidity: Moderate

Body: Light-bodied, low in alcohol

Common descriptors: Aromatic, orange blossom, sage, peach, apricot, grapey, honey

Food pairings: The wine pairs perfectly with dessert, especially fresh fruit, pound cake with fruit, and fruit-based desserts and pastries

NOTABLE PRODUCERS

ASTI
Gancia: The founder, Carlo Gancia, is considered the pioneer of Asti sparkling production

Martini & Rossi: One of the largest and best-known producers

Cinzano: As famous for their marketing posters as they are for their wines; one of the first producers of sparkling wine in the region

MOSCATO D'ASTI
Michele Chiarlo: Credited with convincing consumers outside of Italy that Moscato d'Asti could be a serious wine; the Nivole label is considered a classic example of this style

La Spinetta: Produced the first-ever, single-vineyard bottling of Moscato in Italy, "Bricco Quaglia" Moscato d'Asti

Elio Perrone: Now run by retired champion motocross racer, Stefano Perrone; quickly gaining a reputation as one of the region's rising stars

PRODUCTION NOTES

Asti and Moscato d'Asti share the same DOCG appellation. Both require 100% Moscato Bianco and are rendered sparkling using a version of the tank method; however, the wines are different. Moscato d'Asti is usually made from the best grapes and artisanal crafting often leads to superior quality.

Asti DOCG

- Fully sparkling
- Most Asti DOCG is sweet, but balanced by acidity
- Low alcohol
- Usually not vintage-dated
- Large-scale producers dominate production

Moscato d'Asti DOCG

- Lightly fizzy
- Always sweet, with balancing acidity
- Lower in alcohol
- Vintage-dated
- Artisanal; small-scale producers dominate production

PIEMONTE'S SIGNATURE CUISINE

DESSERT

The wines of Asti are sweet enough to pair with the traditional desserts of Piemonte.

Panna Cotta (cooked cream): This Piemontese delicacy is believed to have originated in the Langhe Hills. It is a pudding-like dessert made of cream, milk, sugar and gelatin.

Zabaione is originally from Piemonte, although it is made throughout Italy. This light custard, believed to have been developed in the Middle Ages, is made by whipping egg yolks with Marsala wine and sugar. Several legends surround its origin; however, the traditional Piemontese story gives credit (and the name) to Saint Baylon.

NOTES

NORTHERN ITALY
White Wines

NORTHERN ITALY WHITE WINES

Map by
Quentin Sadler
WINE
SCHOLAR
GUILD

AUSTRIA

TRENTINO-ALTO ADIGE

Adige

Südtiroll/Alto Adige DOC

Bolzano /Bozen

Friuli Colli Orientali DOC

SLOVENIA

Speck

Trento

Prosciutto di San Daniele

FRIULI VENEZIA GIULIA

Lago di Como

Collio DOC

Lago Maggiore

Lago d'Iseo

Lago di Garda

VENETO

Trieste

Treviso

VALLE D'AOSTA

Valle d'Aosta

Brescia

Vicenza

Venezia

Aosta

LOMBARDIA

Lugana DOC

Soave DOC

Padova

ADRIATIC SEA

PIEMONTE

Milano

Verona

Po

Po

Mantova

Adige

Po

Torino

Ravioli

Po

EMILIA-ROMAGNA

Roero Arneis DOCG

Asti

Gavi DOCG

Parma

Ferrara

Alba

Modena

Vitello Tonnato

Bologna

Ravenna

Genova

Portofino

Rimini

LIGURIA

Italian Riviera

Cinque Terre

GULF OF GENOA

LIGURIAN SEA

TOSCANA

MARCHE

SWITZERLAND

FRANCE

0   50   100 km

0   25   50 miles

N

32

NORTHERN ITALY WHITE WINES:

THE ROAD MAP

Detour to
LUGANA DOC
Loo-GAH-nah

The region: LOMBARDIA/LOMBARDY & VENETO

Lom-bahr-DEE-ah; VEH-neh-toe

The Lugana DOC encompasses the flat plain south of Lake Garda and has territory in both Lombardia and Veneto, but most of the appellation lies in the former. The moderating influence of the lake and the clay-based soils surrounding it combine to produce Lombardia's most distinctive white wines.

③ SOAVE DOC & SOAVE SUPERIORE DOCG
So-AH-veh;
So-AH-veh Soo-pehr-EE-oreh

The region: VENETO
VEH-neh-toe

Veneto is Italy's most productive wine region by volume and has the country's third-largest area under vine (behind Sicilia and Puglia). Even though the majority of the vineyards are planted to white grapes, the region is also famous for its red wines.

Not to miss travel site:
Castello Scaligero, a 10th century castle, overlooks the village of Soave. The castle is connected to medieval walls that surround the ancient part of the town and the vineyards of Soave lie behind it. Dominating the landscape, both castle and hemmed village have become an iconic symbol of the area and its wines.

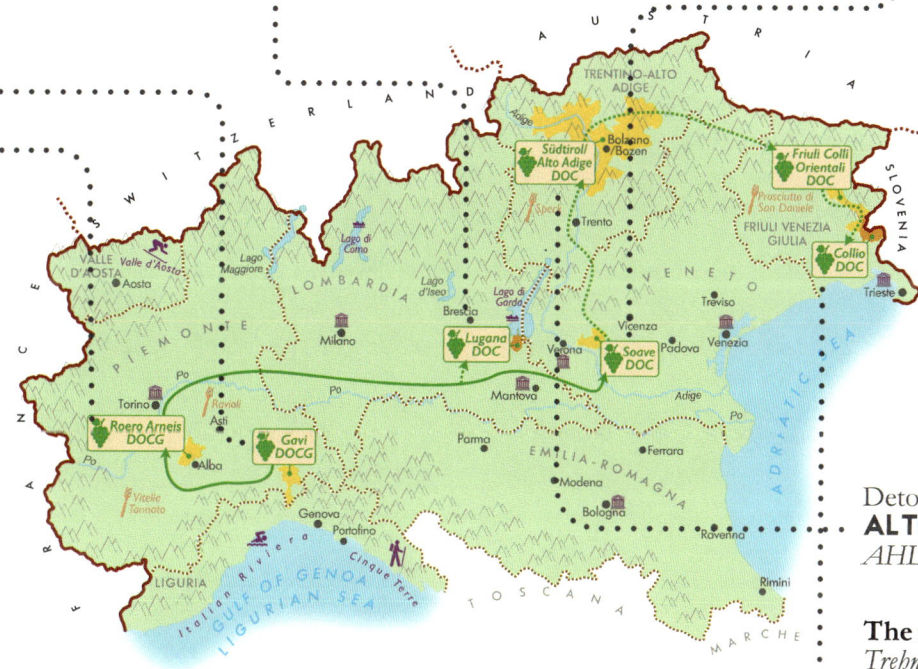

① GAVI DOCG
GAH-vee
② ROERO ARNEIS DOCG
Ro-EH-ro Ahr-NAYS

The region: PIEMONTE/PIEDMONT

Pee-ay-MON-tay

Piemonte is completely landlocked in Italy's northwestern corner and is Italy's second-largest region by size after Sicily. The strategic location, nestled between France, Switzerland and the rest of Italy, made Piemonte an important seat of power for centuries.

Not to miss travel sites:
The hills of Roero and Gavi offer a different kind of tourist draw. The villages are small but the rural scenery is breathtaking. The hills of Piemonte are studded with medieval castles that served as protection for the local populace during the unsettled era of barbarian invasions.

Detour to
COLLIO DOC & FRIULI COLLI ORIENTALI DOC
KOHL-ee-oh; FREE-ewe-lee KOHL-ee Or-ee-en-TAHL-ee

The region: FRIULI VENEZIA GIULIA
FREE-ewe-lee Veh-NEH-zee-ah JOO-lee-ah

A narrow band of hills stretches across the eastern portion of Friuli and into neighboring Slovenia. The best wines of the region are made here. The Alps shield the vineyards from cold northerly winds and help to trap warm Adriatic breezes.

Detour to
ALTO ADIGE DOC
AHL-toh AH-dee-jeh

The region: TRENTINO-ALTO ADIGE
Trehn-TEE-noh AHL-toh AH-dee-jeh

Alto Adige (Südtirol in German) occupies the northern half of the Trentino-Alto Adige region and is Italy's northernmost winegrowing area. Tucked into the northeastern sector of the Italian Alps, Alto Adige borders Austria and Switzerland. The proximity and the historic ties with German-speaking countries have greatly influenced the viticulture and winemaking tradition of the region. Unsurprisingly, several of the principal grapes grown in Alto Adige are Germanic varieties.

GAVI DOCG
Piemonte

In a region that has historically focused on red wines, Gavi, which rose to prominence in the 1960s, has become Piemonte's most well-known dry white wine.

THE PLACE

The Gavi (aka Cortese di Gavi) DOCG lies around the small town of Gavi in southeastern Piemonte near the Ligurian border. Here, Piemonte and Liguria aesthetically merge, creating a landscape of gentle hills and valleys with the majestic Ligurian Apennines as a backdrop.

The vineyards are exclusively planted on hillsides that catch the optimal amount of sun.

The growing conditions are well suited for white wine grapes, thanks to warm summers that are moderated by breezes blowing in from the Ligurian Sea. Cooler temperatures at night slow down ripening, promote aromatic development and enhance the grapes' natural acidity.

GRAPE VARIETIES

Gavi is made from the Cortese (*Kor-TEH-zeh*) grape, which is native to Piemonte. Documents speak of Cortese plantings in this area as far back as the 17th century!

THE FORTRESS OF GAVI

The hilltop Fortress of Gavi (Forte di Gavi) overlooks the eponymous village and is the best-known landmark of the area. This fascinating structure, built originally in the 10th century, serves as a visual timeline for the village's history—through the centuries, successive layers were built upon a medieval base.

GAVI DEL COMUNE DI ...

A wine labeled as "Gavi del Comune di Gavi" means the grapes in that wine were grown within the boundaries of the Gavi village. All of the villages included in the Gavi appellation (plus a few hamlets) are allowed to add their name to the label, provided that is where the grapes were grown.

AGING

Reading about mandating aging periods for wine appellations is tedious at best, so why include them? The aging regimen (or lack thereof) can offer clues about the taste, quality level and ageability of a wine. Though exceptions always exist, aging a white wine before its release for sale allows for acidity to integrate and more time for aroma development.

WINE PROFILE

What makes this wine special: Gavi is regarded as Piemonte's historic white wine and the best appellation for the Cortese grape

Acidity: Pronounced

Body: Light- to medium-bodied, moderate in alcohol

Common descriptors: Crisp, mineral, subtly aromatic, fresh citrus, lemon zest, hints of candied fruit, white flowers and almonds

Food pairings: Appetizers, seafood, light pasta dishes, chicken, salads

PRODUCTION NOTES

- Wines are 100% Cortese
- Gavi can also be produced as Riserva which requires a minimum of one year of aging
- Most Gavi wines are unoaked
- Although most Gavi is still, a sparkling (spumante) version is also produced

CONSORZIO TUTELA DEL GAVI

NOTABLE PRODUCERS

Produttori del Gavi: This cooperative has continued to evolve since its inception in 1951 and now produces some of the region's best value wines

Castellari Bergaglio: Historic producer that makes an excellent range of single-vineyard Gavi highlighting individual terroirs

PIEMONTE'S SIGNATURE CUISINE

PASTA

Ravioli di Gavi
The wines of Gavi are the most famous local product, but the area is also historically known for a culinary specialty, ravioli. Although this renowned stuffed pasta is made throughout Italy and its origin is debated, Gavi has a long-standing tradition for making it. Some even claim that it was invented by the Raviolo family at their tavern in Gavi. Ravioli "al tocco" is a Gavi speciality. The ravioli are filled with meat, cheese and/or vegetables then topped with a meat-based sauce.

CHEESE

Robiola di Roccaverano DOP is a delicate, soft, fresh cheese made in the area around the charming town of Roccaverano in southern Piemonte. It is primarily based on goat's milk but also includes both sheep and cow's milk. The cheese has a distinctive fragrance which is attributed to the aromatic herbs eaten by the grazing animals.

ROERO ARNEIS DOCG
Piemonte

Native to the hills of Roero, the grape, Arneis, makes some of Piemonte's most distinctive and fashionable dry white wines.

THE PLACE

The Roero Arneis DOCG lies in the hills of Roero in the southern part of Piemonte. It is located just to the north of the city of Alba and of Langhe, the land of Barolo and Barbaresco, Piemonte's most famous reds. The Tanaro River separates Roero from Langhe.

The terrain is different on the Roero side of the river. The hills are lower in elevation, but the slopes are steeper. Soils have a higher concentration of sand.

GRAPE VARIETIES

The primary grape in this appellation is the ancient grape Arneis. In the 1960s, Arneis was on the verge of extinction. It was saved by a few prominent producers who bottled varietal versions and rescued the grape from oblivion.

LOCAL DETOUR

The southern and south-central swaths of Piemonte are covered in a series of hills situated close together but distinct from one another. Each set of hills has its own name: The Langhe Hills, The Roero Hills, and The Monferrato Hills.

ROERO DOCG

The Roero DOCG is actually an appellation for white and red wines. While Roero Arneis DOCG is the appellation for varietal Arneis, the Roero DOCG (without the name of the grape) crafts red wines from Nebbiolo. The soil composition and growing conditions are just different enough on the left bank of the Tanaro River that these Nebbiolo wines are consistently lighter in body and more approachable in their youth than the better-known and more powerful wines of Barolo and Barbaresco to the south.

ETYMOLOGY

Two theories exist as to how Arneis got its name. "Arneis" means "little rascal" in the local dialect and is said to reference how challenging the grape is in the vineyard and in the cellar. Others believe the grape was named after a local winegrowing site called "Renesio."

WINE PROFILE

What makes this wine special: Together with Gavi, Roero Arneis is considered Piemonte's signature white wine

Acidity: Soft, low to medium

Body: Medium- to full-bodied, medium in alcohol

Common descriptors: Floral, apple, pear, peach, almond

Food pairings: Appetizers, poultry, fish, salads and other lighter fare

PRODUCTION NOTES

- A minimum 95% Arneis is required; however, wines are usually 100% Arneis

- Roero Arneis can be produced as Riserva which requires a minimum 16 months of aging

- Most wines are unoaked

NOTABLE PRODUCERS

Bruno Giacosa: One of Piemonte's most famous winemakers; he was instrumental in saving Arneis from extinction

Vietti: Credited, alongside Bruno Giacosa, as the savior of Arneis; Vietti began producing dry Arneis in 1967 and the wine has remained an industry icon ever since

Malvirà: Widely recognized as one of the best producers of Arneis; owns the historic Renesio vineyard where the Arneis grape was first identified

Giovanni Almondo: Modern producer of traditional wines focused on terroir-driven bottlings

CONSORZIO TUTELA ROERO

PIEMONTE'S SIGNATURE CUISINE

MEAT

Vitello Tonnato, originally from Piemonte, is now popular throughout Italy. It is a cold dish composed of thin slices of veal, covered with a sauce made with tuna, olive oil, egg yolk and lemon juice (the modern recipe calls for mayonnaise with tuna and capers). Roero Arneis is considered an ideal pairing.

CHEESE

Toma is a soft or semi-hard cheese made from pasteurized cow's milk in Piemonte's Alpine valleys. There are several different Toma cheeses produced in northern Italy, but Toma Piemontese was the first to achieve DOP status.

DETOUR

LUGANA DOC
Lombardia & Veneto

Although Lugana is the favorite libation of tourists relaxing on the beaches of Lake Garda, in recent years the wine has significantly gained in status and reputation both in Italy and abroad. The wines show surprising versatility, offering a range of styles from young and deliciously crisp to complex and surprisingly long-lived.

THE PLACE

Lake Garda is Italy's largest lake and is arguably its most famous. Formed by retreating glaciers, the lake is a popular tourist draw for families and celebrities alike.

The winegrowing area lies to the south of the lake with vineyards stretching from the lakeshore to a string of low, inland hills.

Lugana
D.O.C.
CONSORZIO TUTELA

GRAPE VARIETIES

Lugana is made from the white grape Turbiana *(Tour-BEE-ahna)*, previously known as Trebbiano di Lugana. Scholars and scientists have long debated its relationship to Verdicchio.

Some insist that Turbiana is a clone of Verdicchio (Marche's most important white grape), while others maintain that the two grapes are closely related but distinct varieties. No matter the scientific relationship, it is clear that the wines, in fact, taste different from each other.

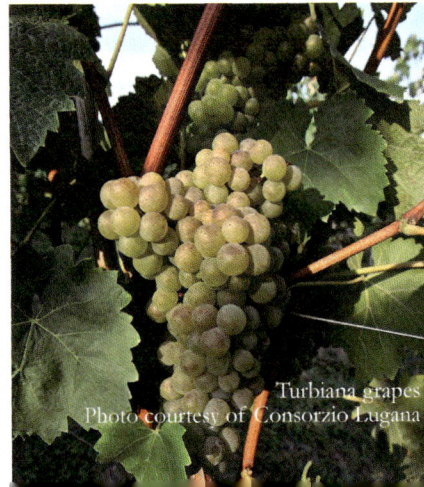

Turbiana grapes
Photo courtesy of Consorzio Lugana

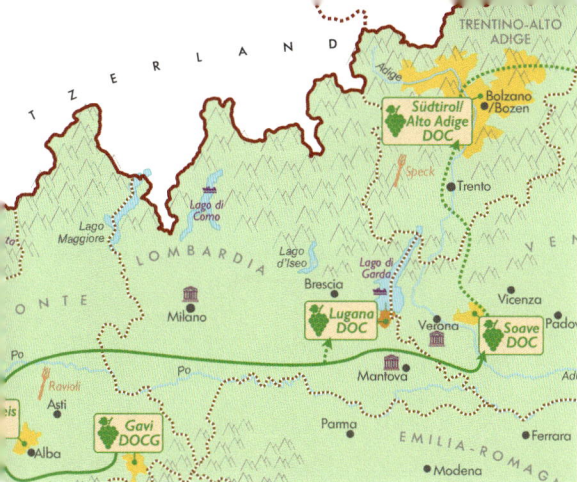

Cà dei Frati: Leading producer that was instrumental in the development of the Lugana DOC

Selva Capuzza: Led by the former Lugana Consortium president, Luca Formentini, the estate is dedicated to the preservation of local cultural traditions and native grape varieties

Zenato: Well-respected, historic producer that has been making wines in Lugana since Sergio Zenato saw the potential of Turbiana in the 1960s

WINE PROFILE

What makes this wine special: Nicknamed the 'liquid gold' of Lake Garda, Lugana still remains a hidden gem in the panorama of Italian white wines. The wines possess crisp acidity, medium- to full-body and aromas of white flowers, citrus, apple, stone fruit, almond and mineral.

What you need to know: The Lugana DOC mostly produces still white wines under one of three categories: Lugana, Lugana Superiore and Lugana Riserva. The entry-level Lugana is a great introduction to the appellation, offering crisp wines with hints of apple and citrus. The Superiore and (especially) the Riserva categories provide a step-up in depth, minerality and aging potential. Lugana pairs well with shellfish and fish, especially lake fish like the trout, perch or lavaret found in Lake Garda.

Torre San Martino della Battaglia

LOMBARDIA & VENETO

SPECIALTY PRODUCTS

Vines are not the only crop cultivated around Lake Garda. The lake is renowned for its olive oil, which has its own appellation—the Olio Garda DOP. The shores are also dotted with citrus groves, particularly on the northwest-ern section of the lake. This area is known as "Riviera dei Limoni" (Coast of Lemons).

SOAVE DOC & SOAVE SUPERIOR DOCG
Veneto

Soave is one of Italy's most well-known and historic white wines. In recent years, an increasing number of quality-conscious producers, in combination with the designation of single cru vineyards, has created new enthusiasm for a wine whose popularity has waxed and waned over time.

THE PLACE

The Soave winegrowing area is northeast of the city of Verona, adjacent to the Valpolicella district (known for its red wines). The hills in the western part of the appellation are steep and limestone based, while the eastern part has lower, volcanic hills. A flat, alluvial plain comprises the southern portion of the area. The most prized and distinctive wines of Soave hail from hillside vineyards within the classico district. The wines here are intense and mineral. This area lies in the center of the production zone and is the heart of the appellation.

It is believed the name of the village of Soave is derived from the ancient Germanic Suavi tribes who established themselves in this area after the fall of the Roman Empire. Interestingly however, the word 'soave' in Italian also means gratifying, or pleasing… which perfectly describes the wines.

SOAVE
CONSORZIO TUTELA

GRAPE VARIETIES

The primary grape of Soave is the native Garganega (*Gar-GAN-ehgah*), which is counted among Italy's most ancient varieties.

CLASSICO

Classico is used on a DOC or DOCG wine label following the name of the appellation (e.g. Soave Classico DOC) to indicate that the grapes were grown in the original, historic winegrowing area.

RECIOTO

The Soave area has been regarded as a quality winegrowing district since at least Roman times. Originally, the wines were sweet and made from air-dried grapes in a process called appassimento. Though no longer the only style of the appellation, these prized sweet wines are still made today and are designated as Recioto di Soave DOCG. Also based on Garganega, the wines show rich texture and complex aromas of flowers, apricot, dried fruit, honey and almond. The sweetness is balanced by fresh acidity and they are ideal dessert or meditation wines.

APPASSIMENTO

Appassimento (drying) is the ancient technique of air-drying grapes. The grapes shrivel as water within the berries evaporates; this concentrates sugars, acids and flavor compounds. The appassimento may be carried out under the sun or indoors in well-ventilated rooms. The process is used in many areas of Italy for both white and red wines.

What makes this wine special: The finest Soave wines are considered among some of Italy's best and longest-lived white wines thanks to their subtle but enticing perfume, depth and minerality

Acidity: Crisp, medium to high

Body: Medium-bodied, medium in alcohol

Common descriptors: Delicate aromas of white flowers, tangerine, lemon, pear, apple, and yellow stone fruit, almond, mineral

Food pairings: Appetizers, mozzarella, seafood (especially squid ink risotto), fish, white meats

NOTABLE PRODUCERS

Pieropan: Historic, quality producer of Soave credited with revitalizing the reputation of the appellation's wines

Inama: Masters of Garganega; known for their expertise in crafting fine wine from the volcanic soils within Soave Classico

Prà: Benchmark producer that has been consistently recognized for the quality and value of their wines

Cantina di Soave: Founded in 1898, this historic producer boasts a broad portfolio of well-made wines that are exported to more than 50 countries

PRODUCTION NOTES

- Many producers use only Garganega; however, Trebbiano di Soave and/or Chardonnay are legal blending partners in small quantities

- There are two appellations within the Soave zone of production. Soave DOC is made from grapes sourced from the entire Soave area including both hills and plains. Soave Superiore DOCG is made with fruit grown only on hillside vineyards in a smaller, delineated area and is released later than Soave DOC. Soave Superiore can also be produced as Riserva which must be aged for one year.

- Soave DOC and Soave Superiore DOCG can be labeled with the designation classico if produced from grapes sourced only in the classico sub-zone

Photo courtesy of Azienda Agricola Graziano Prà

VENETO'S SIGNATURE CUISINE

FISH

Baccalà alla vicentina is one of the most famous specialties of Veneto. Made with dried cod, it was introduced as early as the 15th century by Venetian sailors returning from excursions to Norway. The dried cod soaks for at least 48 hours in cold, fresh water and is later cooked slowly with onions, garlic, parsley, anchovies, olive oil and milk. It is typically served with polenta.

RISOTTO

Risotto al Nero di Seppia (squid ink risotto) is a traditional dish of the Veneto region, particularly in Venice. Squid or cuttlefish are cooked with their ink-sacs, giving the risotto a distinctive black color and rich, briny flavor.

POLENTA

Veneto is responsible for a sizeable portion of Italy's corn production, so it should come as no surprise that polenta is often served on local tables. Polenta is made from cornmeal slowly simmered with water or milk. It can be served with a porridge-like consistency or solidified, sliced and fried or grilled. Either way, this dish is extremely versatile - the number of toppings is endless.

ALTO ADIGE DOC
Trentino-Alto Adige

DETOUR

Alto Adige has gradually become Italy's most dynamic and consistently reliable white wine region. In fact, some of Italy's most distinctive and highest-quality whites come from this German-speaking Alpine region in northeastern Italy.

THE PLACE

The Alto Adige DOC covers all vineyard-viable land in Alto Adige and has several sub-zones based on the growing area. Vines are grown at high elevation on the slopes of the Alps and receive ample sunshine. Warm summer days and cool nights enhance the aromatic development of the grapes and provide for slow and gradual ripening and maturation. Alto Adige boasts some of the highest-elevation vineyards in Europe. Some vineyards reach heights of 4,200 ft/1,300 m above sea level!

What makes this wine special: Alto Adige is known for distinctive, aromatic and elegant white wines. The region's high-elevation vineyards are capable of producing wines with vibrant acidity, pronounced aromatics and significant aging potential.

What you need to know: Varietal Pinot Bianco, Gewürztraminer, Pinot Grigio and Sauvignon Blanc are considered Alto Adige's claim to fame. In particular, Pinot Bianco achieves unmatched levels of quality and aging potential in these hills. Gewürztraminer is considered traditional and is a specialty of the village of Termeno (Tramin). Both Pinot Grigio and Sauvignon Blanc are among Italy's best expression of these varieties.

GRAPE VARIETIES

Alto Adige grows a wide range of international grape varieties. Its primary focus is on white grapes. Pinot Bianco, Pinot Grigio, Chardonnay, Sauvignon Blanc, Gewürztraminer, Sylvaner, Kerner, Riesling and Müller Thurgau, among others, are all grown here with great success.

NOTABLE PRODUCERS

Cantina Terlan: One of the leading winegrowers' cooperatives in Italy; responsible for producing some of Alto Adige's best white wines

Alois Lageder: Progressive and pioneering producer that has incorporated biodynamics and sustainability into their philosophy

Elena Walch: Leading figure in the Alto Adige quality revolution

Abbazia di Novacella: Founded in 1142 by Augustinian monks, this is one of the oldest wineries in the world; renowned for producing wines with a high quality to price ratio

THE NATIVE RED GRAPES OF ALTO ADIGE

Although Alto Adige is mostly known for its white wines, the region is also home to some special, indigenous red grapes used to produce distinctive varietal wines - notably the structured and rich Lagrein and the light and juicy Schiava.

Cooperatives

Cooperatives are an important part of Italy's wine industry and are particularly significant in Alto Adige. Born from difficult economic times, cooperatives were (and are) a way for small growers to pool resources to make and market their wines. Alto Adige has a large number of excellent establishments that are considered not only some of Italy's best cooperatives, but also among the country's best wineries.

43

DETOUR

COLLIO DOC
& FRIULI COLLI ORIENTALI DOC
Friuli Venezia Giulia

No discussion of northern Italian white wine would be complete without mention of the white wines of Friuli Venezia Giulia (shortened as Friuli). This region embraced modern technology in the late 1960s and revolutionized white wine making in Italy. Even today, Friuli remains one of Italy's top white-wine producing regions.

Friuli Colli Orientali DOC

Prosciutto di San Daniele

FRIULI VENEZIA GIULIA

Collio DOC

SLOVENIA

ENETO

Trieste

Treviso

THE PLACE

GRAPE VARIETIES

The eastern hills of Friuli are home to some of the region's most reputed white wines. They are divided into the two hilly wine districts of Collio and Colli Orientali (literally the 'eastern hills'), which also name the two distinctive wine appellations, Collio DOC and Friuli Colli Orientali DOC.

Collio DOC lies along Friuli's eastern border with Slovenia. For the most part, the hills extend from east to west resulting in well-exposed south-facing slopes.

The word "Collio" means "hill" and references the hilly terrain. White grapes hold a solid majority here and are considered some of Italy's best.

The Friuli Colli Orientali DOC lies to the north of Collio. Many hillsides are terraced here; production consists of white wines, well-respected reds and treasured sweet passitos.

Collio DOC and Friuli Colli Orientali DOC share a similar roster of white grape varieties of Italian, Slavic, German and French origin. Friulano, Ribolla Gialla, Malvasia Istriana, Sauvignon Blanc, Chardonnay, Pinot Bianco and Pinot Grigio are common in both appellations. Varieties such as Picolit and Verduzzo, while less common, are often used to produce delicious sweet passito wines.

WINE PROFILE

What makes this wine special: Both Collio DOC and Friuli Colli Orientali DOC make high-quality, terroir-driven varietal and blended white wines from native and international grapes

What you need to know: Friulano is Friuli's signature white wine. This delicately aromatic grape produces wines with citrus and apple fruit laced with notes of hay, herbs and almonds. In Collio DOC and Friuli Colli Orientali DOC, it achieves its consummate expression. Ribolla Gialla is the pride of Collio DOC, where it is produced in white and orange styles. Collio Bianco is a very distinctive and complex blend of local and international grapes. Some of Italy's best Sauvignon Blanc comes from Collio and Friuli Colli Orientali. The elegant and mineral Malvasia Istriana from Collio also merits mention.

NOTABLE PRODUCERS

COLLIO

Schiopetto: Mario Schiopetto, founder of this historic Friuli estate, is considered one of the pioneers of modern Italian white wines

Primosic: Fourth-generation winery that is open to modernization while preserving the teachings of the past

Villa Russiz: This philanthropic estate was instrumental in the importation of French vines into Collio. Its "Foundation Villa Russiz" provides assistance to disadvantaged children. This is a vineyard with multiple missions.

FRIULI COLLI ORIENTALI

Maini: These cult wines are the creation of Enzo Pontoni, arguably Italy's finest white winemaker; production is only 700 cases per year

Livio Felluga: Instrumental in the restoration of abandoned vineyards and the re-establishment of winemaking traditions in Friuli following the end of the Second World War

THE UNIQUE WINE STYLES OF FRIULI

Super Whites

The term "super whites" is sometimes used to refer to high-quality blends incorporating traditional Friulian grapes (notably Friulano, Ribolla Gialla, Malvasia Istriana and Picolit) with international varieties such as Sauvignon Blanc, Chardonnay, Riesling and Pinot Bianco, among others. The term references the famed "Super Tuscans" of Toscana which also feature international grapes. These "super whites" are typically labeled as Collio Bianco DOC, Friuli Colli Orientali Bianco DOC or Venezia Giulia IGT.

Orange Wines

Friuli, particularly Collio, is home to some of the most iconic orange wines made today, largely thanks to the visionary winemaker Josko Gravner. Gravner began experimenting with orange winemaking techniques in the 1990s and inspired a generation of winemakers to follow suit.

White wines are generally made with very little (if any) contact between the must and grape skins and are typically aged less than red wines before bottling. However, it is possible to use red winemaking techniques on white grapes. Skin contact, extended maceration and extra aging time result in golden- or amber-hued wines with strong phenolic character and (often) oxidative notes. Wines made in this fashion are referred to as "orange wines." Though they may seem strange to the uninitiated, they are often some of the most interesting and thought-provoking wines on the market.

NOTES

NORTHERN ITALY
Red Wines

NORTHERN ITALY RED WINES

0   50   100 km
0   25   50 miles

N

AUSTRIA

SWITZERLAND

TRENTINO-ALTO ADIGE

FRIULI VENEZIA GIULIA

SLOVENIA

Adige

Bolzano/Bozen

VALLE D'AOSTA

Valle d'Aosta

Aosta

Lago di Como

Lago Maggiore

Trento

Asiago

V E N E T O

P I E M O N T E

Lago d'Iseo

Lago di Garda

Valpolicella
**

Treviso

Trieste

Tajarin

Milano

Brescia

Vicenza

Venezia

Padova

**Valpolicella zone includes:
Valpolicella DOC
Valpolicella Ripasso DOC
Recioto della Valpolicella DOCG
Amarone della Valpolicella DOCG

Barbera d'Asti DOCG

Torino

Asti

Po

L O M B A R D I A

Verona

Pandoro

Mantova

Adige

Po

Dolcetto d'Alba DOC

Barbaresco DOCG

Barolo DOCG

Parma

E M I L I A - R O M A G N A

Ferrara

Po

A D R I A T I C   S E A

Po

White Truffles

Modena

Bologna

Genova

Portofino

Cinque Terre

Ravenna

LIGURIA

Italian Riviera

GULF OF GENOA

LIGURIAN SEA

T O S C A N A

Rimini

M A R C H E

F R A N C E

Map by
Quentin Sadler
WINE
SCHOLAR
GUILD

① DOLCETTO D'ALBA DOC
Dol-CHEHT-toe DAHL-bah

② BARBERA D'ASTI DOCG
Bahr-BEH-rah DAHS-tee

③ BAROLO DOCG
Bah-ROH-lo

BARBARESCO DOCG
Bar-BAH-rehs-co

NORTHERN ITALY RED WINES:

THE ROAD MAP

④ VALPOLICELLA DOC
Vahl-po-lee-CHEHL-lah

AMARONE DELLA VALPOLICELLA DOCG
Ah-mah-RO-neh DEHL-lah Vahl-po-lee-CHEHL-lah

The region: **VENETO**
VEH-neh-toe

Veneto's best-known wine districts are found in a hilly band at the foothills of the Venetian Prealps. The elevation of the hills keeps temperatures cool at night (crucial to preserving acidity in the grapes) but the hills also trap warm air drifting northward from the Adriatic Sea. The combination is perfect for grape-growing.

Not to miss travel sites:

Although less known than world-famous cities like Venice or Florence, Verona is without doubt a jewel among Italy's art-rich cities. Thanks to its well-preserved architecture and monuments dating back to the Roman, medieval and Renaissance periods, the city of Verona has been named a UNESCO World Heritage Site.

Verona is also well known in the wine world as the city that hosts Vinitaly, Italy's principal wine fair and one of the world's largest international wine events. It takes place every year in April.

The region: **PIEMONTE/PIEDMONT**
Pee-ay-MON-tay

Piemonte has the most DOCs and DOCGs of any Italian region and has no IGTs. More than 80% of Piemonte's wine comes from the southern and eastern sections of the region, where every inch of suitable land is planted to the vine.

Not to miss travel site: The town of Alba, located between Barolo and Barbaresco, is considered the virtual capital of the Langhe wine district. It is also famous for being its culinary center and not by chance—a paradise for lovers of fine food and wine.

Valpolicella ******

**Valpolicella zone includes:
Valpolicella DOC
Valpolicella Ripasso DOC
Recioto della Valpolicella DOCG
Amarone della Valpolicella DOCG

Barbera d'Asti DOCG

Dolcetto d'Alba DOC

Barbaresco DOCG

Barolo DOCG

DOLCETTO D'ALBA DOC
Piemonte

There are several other Dolcetto appellations in Piemonte; however, the most highly regarded are Dogliani DOCG and Dolcetto di Diano d'Alba DOCG.

Dogliani DOCG: The village of Dogliani is considered by many to be the capital of Dolcetto; its Dolcettos carry a prestigious pedigree. The Dogliani zone of production is situated just to the south of the Barolo appellation and crafts what is considered the most concentrated and age-worthy Dolcetto in Piemonte.

Dolcetto di Diano d'Alba DOCG: The small village of Diano d'Alba, just to the south of Alba, has a long-standing tradition for the production of Dolcetto. The best wines come from an official list of 76 cru vineyards, locally known as sörì (superior vineyard sites).

Dolcetto is the traditional everyday red wine of Piemonte. It is produced across most of the region, but strongholds are the hills of Langhe and Monferrato. The Dolcetto d'Alba DOC is Piemonte's largest Dolcetto appellation by volume. It is also the most diverse in style, producing wines that range from light and fruity to fuller-bodied and structured.

THE PLACE

Dolcetto d'Alba DOC is made near Alba (as the name of the appellation implies), in the northern part of Langhe. It overlaps the winegrowing areas of Barolo and Barbaresco entirely. In fact, many Barolo and Barbaresco producers also make Dolcetto d'Alba.

In Langhe, Dolcetto is well respected for its ability to thrive in cooler and less ideal sites, where Nebbiolo and Barbera would not be able to ripen successfully.

GRAPE VARIETIES

Dolcetto is one of Piemonte's historic and most widely planted grapes. The name Dolcetto translates to "little sweet one;" however, this moniker does not refer to the wine, but rather to the grapes. Ripe Dolcetto grapes have moderate acidity and taste deliciously sweet, so much so that they can be eaten as table grapes.

Although Dolcetto usually ranks third in importance and prestige after Nebbiolo and Barbera, it nonetheless plays a critical role among producers. Dolcetto ripens significantly earlier than the other two grapes and the wines are released to market in advance of the other two, making it a welcome source of income while Nebbiolo and Barbera wines mature in cellar.

WINE PROFILE

What makes this wine special: Dolcetto makes deeply colored, soft, succulent, mouth-filling wines with fragrant, vinous aromas and a signature bitter-sweet finish

Acidity: Moderate

Tannin: Fairly tannic

Body: Medium- to full-bodied, medium in alcohol

Common descriptors: Cherry, damson plum, grape, almond, licorice

Food pairings: Appetizers, carpaccio, charcuterie, legumes, soup, pasta, young to medium-aged cheeses

PRODUCTION NOTES

- Dolcetto d'Alba is made from 100% Dolcetto

- Dolcetto d'Alba designated as Superiore must undergo at least one year of aging

- Most Dolcettos are released early and meant for early drinking; however, some producers make bigger, more structured, oak-aged versions

NOTABLE PRODUCERS

DOLCETTO D'ALBA

Marcarini: Known for the centenarian, ungrafted vines that contibute to the famous Boschi di Berri bottling

Vajra: Remains committted to showcasing the best Dolcetto has to offer by planting the vines in locations ideally suited to the variety

Brovia: Famous Barolo producer that also makes a delicious Dolcetto

DOGLIANI

Poderi Einaudi: Founded in 1897 by Luigi Einaudi, a gentleman who became Italy's first President in 1948; Poderi Einaudi was the first estate in the Langhe to combat phylloxera by grafting their vines onto American rootstocks

DOLCETTO DI DIANO D'ALBA

Alario Claudio: Producer who transformed the family estate by focusing on Dolcetto; produces some of the region's finest Dolcettos

PIEMONTE'S SIGNATURE CUISINE

TRUFFLES

The town of Alba, in addition to being the de facto capital of the Langhe wine district, is also famous for being its culinary center and the epicenter for Piemonte's most sought-after treasure… white truffles. Their intense and unique aroma makes them the most prized (and expensive) of the truffles. Every year in October and November, Alba is proudly home to the International White Truffle Fair where it is possible to taste and buy the 'white gold' of Piemonte.

PASTA

Agnolotti are hand-made, square-shaped ravioli, commonly found throughout Piemonte. They are traditionally stuffed with leftover roasted meat. In Langhe and Monferrato, there is a well-known smaller and rectangular variant known as 'Agnolotti del Plin'. The name comes from the pinch (plin) as they are pinched to take their particular shape. Agnolotti del Plin and Dolcetto is considered a classic pairing in Langhe.

Barbera d'Asti is not the only highly regarded appellation for Barbera. In fact, Piemonte boasts several Barberas from different winegrowing areas. Among the most respected are:

Barbera d'Alba DOC: This is the second-largest Barbera appellation by volume and it produces some of Piemonte's finest. The DOC encompasses the territories of Barolo, Barbaresco, and Roero. Thus, it should come as no surprise that many producers of Barolo, Barbaresco and Roero produce a Barbera d'Alba as well. To compare with Barbera d'Asti, Barbera d'Alba tends to be a bigger, richer wine with slightly less pronounced acidity.

Nizza DOCG: The area around the town of Nizza was once part of the Barbera d'Asti DOCG but was elevated to its own appellation thanks to its long-standing reputation for producing high-quality Barbera. In comparison to other Barbera DOC/Gs, overall production rules are the most stringent, especially with regard to yields and aging requirements.

Barbera del Monferrato DOC and Barbera del Monferrato Superiore DOCG: Barbera del Monferrato hails from Monferrato and is usually lighter than wine from other Barbera appellations; however, the Superiore DOCG wines tend to be more structured and fuller-bodied.

BARBERA D'ASTI DOCG
Piemonte

Barbera is cultivated in several areas of Piemonte; however, the area around Asti, among the hills of Monferrato, is considered its true heartland. Here, the grape is able to achieve its highest expression combining juicy fruitiness, vibrant acidity and minerality.

THE PLACE

Barbera d'Asti DOCG is Piemonte's second-largest wine appellation by volume and size. (Asti DOCG is the largest). Barbera d'Asti covers a large part of the Monferrato area around the city of Asti.

The beauty of the vine-covered, rolling hills of Monferrato, together with the stunning vineyard landscapes of Langhe and Roero, has earned this area of Piemonte a listing among the UNESCO World Heritage Sites.

In Langhe, Nebbiolo is the primary focus and is planted on the best sites; Barbera takes a secondary role in production and vineyard location. In Monferrato, Barbera takes the lead and has always been given the best vineyard sites.

GRAPE VARIETIES

Barbera is one of the traditional and most important grapes of Piemonte. It is by far Piemonte's most widely planted variety. Barbera's versatility makes it suitable for a wide range of styles from light, fruity everyday wines to serious, structured, oak-aged bottlings.

What makes this wine special: Barbera d'Asti embodies the archetypal expression of Barbera—a wine hallmarked by deep color, crisp acidity, soft tannins and bright cherry fruit. Ironically, Barbera d'Asti also pioneered the aging of Barbera in new oak, showcasing the grape's natural affinity for maturation in wood.

Acidity: Very high

Tannin: Low to moderate

Body: Light- to full-bodied with medium to high alcohol

Common descriptors: Cherry, raspberry, blueberry, blackberry, underbrush, mineral. Oaked versions show hints of vanilla, chocolate and sweet spices.

Food pairings: Pasta, pizza, hamburgers, chili, charcuterie, rich and fatty dishes

PRODUCTION NOTES

- Barbera d'Asti is made from at least 90% Barbera, although most examples are 100% Barbera

- Barbera d'Asti Superiore is made from selected, riper grapes and must be aged for at least 14 months, including a six-month stint in oak; this category includes some of Piemonte's best and longest-lived Barbera wines

NOTABLE PRODUCERS

BARBERA D'ASTI

Braida: Pioneer producer of high-quality, oak-aged Barbera; the 1985 release of Bricco dell'Uccellone became a landmark wine, proving to the world that Barbera was capable of making serious, cellar-worthy wines

Vietti: Founder Alfredo Currado persisted in planting Barbera in top vineyard sites because he was convinced the grape was capable of excellence when given the chance; for Currado, Barbera was a labor of love

BARBERA D'ALBA

Massolino: Famous Barolo producer that crafts expertly made, easy-drinking examples of Barbera

BARBERA DEL MONFERRATO SUPERIORE

Iuli: Winemaker Fabrizio Iuli practices organic viticulture and proclaims his love for, and dedication to, Barbera by referring to himself as a "barberista"

PIEMONTE'S SIGNATURE CUISINE

SAUCE

Bagna Cauda, a sauce made of olive oil, garlic and anchovies, is a traditional Piemontese gastronomic specialty. The name means "hot bath"—a fitting moniker for a hot dipping sauce for vegetables. It is typically served during autumn and winter as an antipasto and presented in a terracotta bowl atop a tea-light food warmer. Barbera is considered the ideal accompaniment. It is traditionally consumed by farmers to celebrate the end of the harvest and is meant to be enjoyed in company.

PASTA

Tajarin is hand-made pasta belonging to the culinary tradition of Piemonte, particularly in Langhe. Tajarin is a very thin tagliatelle with a deep yellow color and a rich taste due to the large number of egg yolks used in the recipe. Cooked tajarin is typically topped with roasted-meat sauce or tossed with butter and white truffles.

BAROLO DOCG &
BARBARESCO DOCG
Piemonte

Undeniably, Barolo and Barbaresco stand among Italy's greatest red wines and are considered the most revered expression of the noble, red Nebbiolo grape. Traditionally made from a blend of vineyards and villages, they are now increasingly bottled as single-vineyard crus.

THE PLACE

The villages of Barolo & Barbaresco lie in the hilly area of Langhe in the southern part of Piemonte, on the right bank of the Tanaro River. These hills are characterized by hazelnut groves, castles, picturesque villages and vineyards. In fact, Langhe boasts one of Italy's highest concentrations of vineyards and wineries.

These hills, together with the hills of Roero and Monferrato, have been included in the list of UNESCO World Heritage Sites with the official name of "Vineyard Landscape of Piedmont: Langhe, Roero and Monferrato" in recognition of the remarkable scenery of this historic and prestigious winegrowing area.

The winegrowing area of Barolo DOCG lies southwest of the town of Alba and includes eleven villages, including the village of Barolo itself.

The Barbaresco DOCG is northeast of Alba. By comparison, its winegrowing area is smaller (as is its production volume). It includes only three villages, among which is the village of Barbaresco itself.

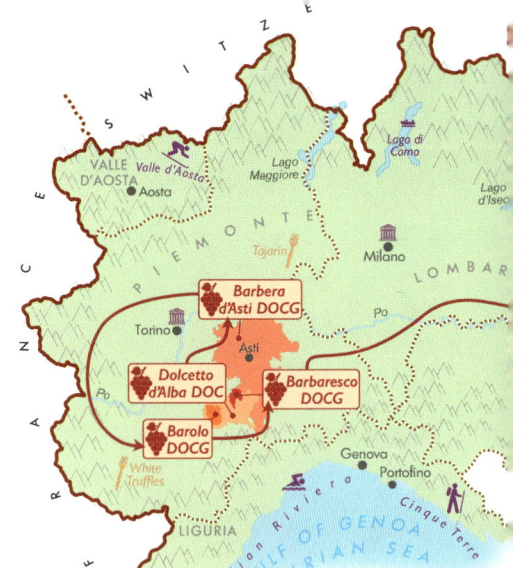

GRAPE VARIETIES

Nebbiolo (*Naeb-bee-OH-loh*) is one of Italy's finest grape varieties as well as one of the oldest. It has been mentioned in written record since the 13th century! Nebbiolo is often compared to Pinot Noir thanks to its ability to manifest the nuances of different terroirs.

BAROLO VS BARBARESCO

In the 19th century, Barolo was known as "re dei vini e vino dei re" (king of wines and wine of kings) and is still considered the greatest, most powerful and longest-lived expression of the Nebbiolo grape. Although there are exceptions, Barbaresco is traditionally considered slightly more approachable, lighter and earlier-maturing than Barolo, though it still shows plenty of acidity, substantial tannin and aging potential.

ETYMOLOGY

The name Nebbiolo comes from the Italian word 'nebbia' (fog). Two main theories exist as to how the grape got this name. Some say it refers to the thick waxy bloom that covers the berries, while others attribute the name to the fog that blankets the vineyards in the autumn.

What makes this wine special: In Barolo and Barbaresco, Nebbiolo is able to achieve an alluring perfume, impressive structure, great complexity and has long aging potential. These attributes place these wines among the finest in the world.

Acidity: Very high

Tannin: Very high

Body: Full-bodied and high in alcohol

Common descriptors: Pale ruby-garnet in color, turning brick-orange with bottle age. In its youth, the wine is floral (rose, violet) and earthy, with aromas of licorice and red cherry. As it matures, the wine softens and develops increasingly complex notes of withered flowers, dried red fruit, leather and truffles.

Food pairings: Barolo and Barbaresco pair well with white truffle- or mushroom-based pastas and risottos and expertly complement rich roasts, braised meat, game and aged cheese

PRODUCTION NOTES

- Barolo and Barbaresco are made entirely from Nebbiolo. Traditionally, both are aged for an extensive period of time and greatly benefit from bottle aging.

- Barolo requires a minimum 38 months of aging with at least 18 months in oak barrels. Barolo Riserva must be aged for a minimum of 62 months with the same minimum period in oak.

- Barbaresco requires a minimum 26 months of aging with at least nine months in oak barrels. Barbaresco Riserva must be aged for a minimum of 50 months with the same minimum period in oak.

NOTABLE PRODUCERS

BAROLO

Giacomo Conterno: Iconic Barolo producer, his legendary Barolo Riserva Monfortino is considered among Italy's best wines

Bartolo Mascarello: Historic, traditionalist; a top-quality Barolo producer

Giuseppe Mascarello: Top-quality producer with highly regarded Barolo from the Monprivato Cru vineyard

Giacomo Fennochio: Known as a non-interventionist producer who crafts wines of excellence; Fennochio Barolos are often cited as some of the region's "best value" wines

Rivetto: Currently the only Demeter certified biodynamic winery in the Barolo region; diversity and sustainability are hallmarks of this estate

BARBARESCO

Gaja (also Barolo): Angelo Gaja is credited with transforming Barbaresco into a world-class wine with cult status; he was also instrumental in raising the image of Italian wines in general

Produttori del Barbaresco: This highly admired cooperative was one of Italy's first; it is often described as Italy's best cooperative

Bruno Giacosa: Legendary, traditionalist producer of Barbaresco and Barolo

Roagna: Known for their dedication to biodiversity and old-vines, Roagna produces complex, structured wines with incredible purity of fruit

Rizzi: The Dellapiana family owns some of the best vineyards in Barbaresco, notably the famous Rizzi hill portion of the larger Rizzi Cru; producing some of the best value Barbarescos on the market.

PIEMONTE'S SIGNATURE CUISINE

MEAT

Brasato al Barolo is a classic Langhe dish made with beef from Fassona cattle, the renowned, local breed. The meat is marinated in Barolo wine for at least half a day, then stewed very slowly with more Barolo wine, vegetables, aromatic herbs and spices. This long cooking process allows the meat to absorb the flavors of the ingredients and become very tender. It is often served with polenta and is, of course, paired with Barolo wine at the table.

AMARONE DELLA VALPOLICELLA DOCG
Veneto

Besides Amarone, the district of Valpolicella produces three other distinct wines, each with its own appellation:

Valpolicella DOC: This wine's light, zippy, juicy style, redolent of sour cherry, put the Valpolicella region on the map. Valpolicella Superiore DOC is a fuller, more structured style within this category.

Valpolicella Ripasso DOC: Ripasso is a local production technique in which freshly vinified Valpolicella wine is poured through the pomace (vinacce) of Amarone to initiate a re-fermentation. This produces a fuller, richer style of Valpolicella that is stylistically between the light Valpolicella and the fuller Amarone.

Recioto della Valpolicella DOCG: This is an ancient, sweet passito red wine made from air-dried grapes. This is an intense and complex wine with aromas of ripe berry, dried fruit, maraschino cherry and chocolate. The sweetness is nicely balanced by the combined effect of acidity and tannin.

ETYMOLOGY

The area of Valpolicella has been known for wine production since at least Roman times. The historic importance of viticulture in Valpolicella is evident by the name itself. It is believed the name Valpolicella derives from Latin, "Vallis-polis-cellae," meaning the "valley of the many cellars."

Valpolicella is one of northern Italy's historic and most important red wine districts. Once only known for the lively and easy-drinking red wines of the Valpolicella DOC, its reputation has been enriched by the intense and powerful Amarone della Valpolicella DOCG, considered among Italy's true classics.

**Valpolicella zone includes:
Valpolicella DOC
Valpolicella Ripasso DOC
Recioto della Valpolicella DOCG
Amarone della Valpolicella DOCG

THE PLACE

The **Valpolicella district** lies in the western part of Veneto, some 20 miles to the east of Lake Garda and just to the north of the city of Verona.

Topographically, Valpolicella is composed of a series of hills and valleys situated in a north-south orientation, radiating out like fingers from the foothills of Monti Lessini. Vines grow in a pastoral setting characterized by cypresses, woods, olive and cherry trees.

GRAPE VARIETIES

The wines of Valpolicella are primarily based on the red Corvina (Kor-VEE-nah) grape blended with other local red varieties such as Corvinone (Kor-vee-NOE-nae) and Rondinella (Ron-dee-NEL-la).

What makes this wine special: Amarone is a powerful, dry red wine made by appassimento (air-drying the grapes before fermentation.) It is Valpolicella's most prestigious and expensive wine.

Acidity: Lively, fresh acidity balances the wine's typical richness

Tannin: Soft, round, velvety tannins

Body: Opulent, full in body, rich texture and very high in alcohol

Common descriptors: Intense and complex aromas of ripe berry and dried fruit, chocolate, rum, tobacco, licorice and sweet spices

Food pairings: Ideally paired with rich, robust dishes including game, roasts, braised meat and aged cheeses

PRODUCTION NOTES

- The red wines of Valpolicella are principally produced from a blend of Corvina, Corvinone and small proportions of Rondinella. By law, Corvina or Corvinone account for 45-95% of the blend, although Corvina is usually the dominant grape in the blend.

- Amarone and Recioto both undergo appassimento. Grapes are left to air-dry for three to four months. During this period, they lose 35-50% of their water content! This process concentrates the sugar and flavor in the grapes and results in blockbuster wines.

- Amarone must have a minimum alcohol content of 14% abv and must be aged for at least two years. It can also be produced as Riserva which must be aged for no less than four years.

NOTABLE PRODUCERS

Bertani: One of the historic producers of Valpolicella and among the first to purposefully ferment a Recioto to dryness to create Amarone (1950s)

Dal Forno: Renowned and remarkable producer; known for impressive, highly concentrated, powerful wines

Quintarelli: Iconic, traditional producer who has served as inspiration for several other high-quality producers in Valpolicella

Tedeschi: Historic producer that respects tradition while incorporating modern techniques and technology

Zenato: Modern producer committed to making consistent wines of high-quality at affordable price points

VENETO'S SIGNATURE CUISINE

RISOTTO

Risotto all'Amarone is one of the most popular dishes within Veronese culinary tradition. It is considered the signature dish of many renowned restaurants in the city. This dark risotto is made using two of the best products of the area: Amarone della Valpolicella and the Vialone Nano IGP rice. Vialone Nano is a famous type of rice cultivated south of Verona. It is considered one of the best varieties of rice for making risotto.

NOTES

CENTRAL ITALY
White Wines

EMILIA-ROMAGNA

SAN MARINO

Pesaro

Lucca
Firenze
Pisa
Urbino
Ancona

Livorno
Jesi

San Gimignano
Arno
Arezzo
Gubbio

Siena
Cortona
Perugia
MARCHE

Bolgheri
Montepulciano
Assisi

TOSCANA
Montalcino
Lago Trasimeno
Ascoli Piceno

Montefalco

Grosseto
Orvieto
Norcia
ABRUZZO

Elba
UMBRIA
Corno Grande
9,554 ft/2,912 m
Pescara

Lago di Bolsena
GRAN SASSO
Rocca Calascio

L'Aquila

Tarquinia
LAZIO

Lago di Bracciano
Monte Amaro
9,161 ft/2,793 m

Tiber
Cerveteri
Tivoli

Roma
Frascati

Ostia Antica
MOLISE

Campobasso

ADRIATIC SEA

TYRRHENIAN SEA

A P E N N I N E S

P U G L I A

C A M P A N I A

N

| 0 | 50 | 100 km |
| 0 | 25 | 50 miles |

Map by
Quentin Sadler
WINE
SCHOLAR
GUILD

CENTRAL ITALY

The central part of Italy gave birth to the Renaissance. But as P.J. O'Rourke, the American political satirist and journalist, commented, "Not much was really invented during the Renaissance, if you don't count modern civilization."

The Renaissance advancements in art, literature, science and politics went on to change the very fabric of the world and to shape modern society… and central Italy was the epicenter.

Central Italy is also the birthplace of the modern Italian language. So many important and influential writers emerged from Toscana during the Renaissance that the Tuscan form of Italian became the gold standard of the language for the entire country.

The roots of Italian culture, both Roman and Etruscan are located here. The eternal city of Roma and Vatican City (which encompasses the Holy See of the Catholic Church) lie at the heart of central Italy and were the epicenters of political and religious power in Italy, in Europe and beyond… for centuries.

Much of central Italy has been shaped by a feudal system of agriculture known as mezzadria (sharecropping). Under this system, wealthy estate owners controlled much of the agricultural land. Essential to the system were the mezzadri (families that lived on and farmed the land). In return for the privilege of living on and farming the land, the mezzadri were required to turn over half of everything they produced to the estate-owner (padrone). The entire economy was agriculture based.

Mezzadria was abolished in the 1960s. Farmers then left the countryside and moved to urban centres in search of jobs. Farmland was abandoned and properties fell into disrepair. Investors began to snap up inexpensive properties and revitalize them. Although agriculture remains a major economic activity, investment has encouraged tourism, small businesses and cottage industries specializing in hand-crafted items such as ceramic tiles and shoes.

Much of central Italy is comprised of beautiful hillsides dotted with olive groves, vineyards and rolling fields of wheat, barley and other crops. Rich meaty sauces are a typical accompaniment to fresh and dried pastas. The many forests provide an abundance of game, such as the famous wild boars (cinghiale) that terrorize grape growers with their insatiable appetite for juicy, ripe grapes. Toscana is particularly well known for its bistecca (steak) from the famous Chianina cattle.

Umbria, the "Green Heart of Italy" is renowned for the quality of its cured meats, but its most famous foodstuffs are porcini mushrooms and truffles. The cuisine of Lazio is often defined by the cuisine of Roma. The area is home to some of Italy's best (and somewhat flatter) agricultural land and the quality of the produce is second to none. Many dishes are simply prepared to highlight the flavor of these fresh ingredients. Artichokes, pasta with aglio e olio (garlic and olive oil) and veal tripe stewed with tomatoes are some of its more famous offerings.

Abruzzo and Marche were historically poorer regions. Here, the typical diet was much more vegetable-focused, particularly for those living inland. Lamb was often the meat of choice, but its inclusion at the table was largely saved for celebratory meals. Those living near the coast supplemented their meals with seafood.

Central Italy is home to some of Italy's most iconic landscapes and attractions. The cities of Roma and Firenze throng with tourists eager to experience all the cities have to offer. However, Central Italy is also home to some of the country's most beautiful and less-travelled areas. Visitors willing to get off the beaten path and explore more of Central Italy won't be sorry they took the road less travelled!

CENTRAL ITALY WHITE WINES

EMILIA-ROMAGNA

SAN MARINO

Vin Santo di Carmignano DOC

Vin Santo del Chianti Classico DOC

Verdicchio dei Castelli di Jesi DOC

Vin Santo del Chianti DOC

Vernaccia di San Gimignano DOCG

Vin Santo di Montepulciano DOC

Orvieto DOC

Lucca
Cantucci
Firenze
Pisa
Livorno
San Gimignano
Saffron
Siena
Arezzo
Arno
Cortona
Montepulciano
Lago Trasimeno
Montalcino
Val d'Orcia
Orvieto
Elba
Maremma
Lago di Bolsena

TOSCANA

LAZIO

Tarquinia

Lago di Bracciano
Tiber
Cerveteri
Tivoli
Roma
Frascati
Ostia Antica

TYRRHENIAN SEA

Pesaro
Urbino
Ancona
Jesi
Conero
Gubbio
Grotte di Frasassi
Perugia
Assisi
Olives all'Ascolana
Montefalco
Norcia
Ascoli Piceno

MARCHE

UMBRIA

Olive Oil

Gran Sasso
Pescara
L'Aquila
Rocca Calascio
Saffron

ABRUZZO

ADRIATIC SEA

MOLISE

Campobasso

PUGLIA

CAMPANIA

0   50   100 km
0   25   50 miles

N

CENTRAL ITALY WHITE WINES:

THE ROAD MAP

1 VERNACCIA DI SAN GIMIGNANO DOCG
Vehr-NAHT-chee-ah dee SAHN Gee-mee-NYAH-noh

The region: TOSCANA/TUSCANY
Tohs-CAH-nah
The rolling hills of Toscana are one of Italy's most iconic landscapes; the romance of its picturesque countryside looms large in the hearts of many.

Not to miss travel sites:
Quite a few of Italy's most famous cities and most revered tourist sites are found in Toscana. Firenze (Florence) has been a center of political and artistic life for centuries.

The Piazza della Signoria, an L-shaped square in front of the Palazzo Vecchio (the town hall), is considered the "beating heart" of Firenze. The Piazza leads to the Uffizi Gallery, housing one of Italy's greatest collections of Renaissance art. The Accademia Gallery, is home to Michelangelo's masterpiece, "David."

The Ponte Vecchio (old bridge) is an important medieval arched bridge open in the center, with shops lining both sides. It is one of the most recognizable symbols of Firenze.

2 ORVIETO DOC
Or-vee-EH-toe

The region: UMBRIA & LAZIO
Oom-BREE-ah; LA-tsee-oh
Umbria is the only landlocked region in central Italy and is one of the country's least populated. This is Italy's hilliest region. The hills are covered by thick forests, alternating with pastures, olive groves and vineyards.

Not to miss travel sites:
Umbria's hills are dotted with dozens of medieval walled towns, each with a historic and beautiful central square, an ancient town hall, and a striking cathedral.

Orvieto is perhaps Umbria's most beautiful medieval town. It is well known for the Duomo di Orvieto, a magnificent cathedral with a stunning mosaic façade and home to a marble Pietà sculpture. The town is built atop volcanic rock and has an interesting network of caves and tunnels beneath it that were built by the Etruscans. These caves are ideal for winemaking and are still used for that purpose today.

3 VERDICCHIO DEI CASTELLI DI JESI DOC
Vehr-DEEK-kyo Day Kahs-TEHL-lee Dee YEH-zee

The region: MARCHE
MAR-keh
The Marche region exhibits the quintessential central Italian landscape and many parts of it remain untouched by urban development. The Central Apennines run along the western portion of the region, providing a breath-taking backdrop to the band of gentle hills that fade into the Adriatic Sea.

Not to miss travel sites:
Urbino is a medieval, walled city designated as a UNESCO World Heritage Site for a remarkable legacy of Renaissance art, architecture and culture. The hillside town is best-known for the Palazzo Ducale, a 15th century architectural masterpiece.

The mountains surrounding the town of Ascoli Piceno supplied medieval builders with an ample supply of travertine marble. The central square, Piazza del Popolo and its adjacent buildings, are all made of this grey-hued stone, resulting in a stunning display of medieval architecture that has been well preserved for centuries.

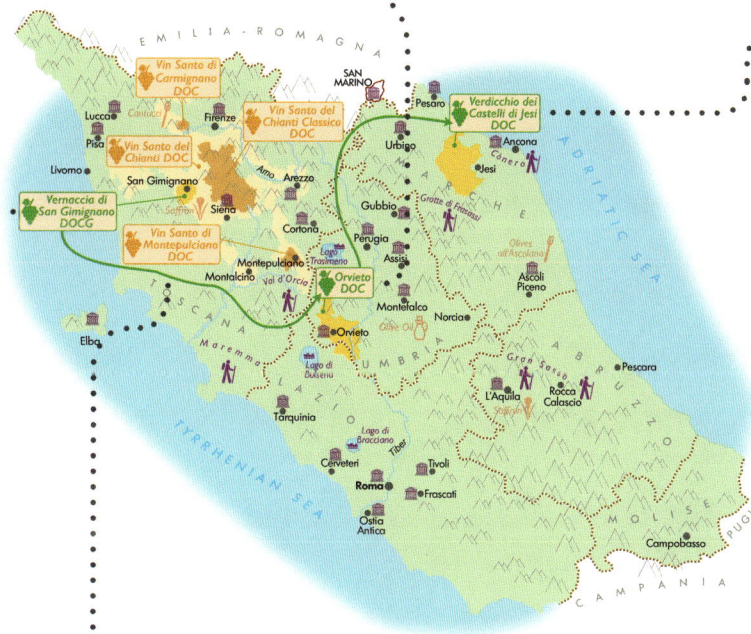

Detour to
VIN SANTO DOCs
Vin SAHN-toe

The region: TOSCANA/TUSCANY

Toscana has the second-largest number of DOCs and DOCGs (after Piemonte). A large percentage of its wines are exported, making Toscana (arguably) Italy's best internationally known wine region.

VERNACCIA DI SAN GIMIGNANO DOCG
Toscana

San Gimignano is one of Toscana's most picturesque villages and its wines are among Italy's most historically renowned and respected. In recognition of its cultural significance, Vernaccia di San Gimignano was the first Italian wine to be awarded DOC status in 1966. Its status was upgraded to DOCG in 1993 and it remains the only DOCG for white wine in Toscana.

ETYMOLOGY

The name "Vernaccia" is believed to be derived from the Latin word "vernaculus" meaning "indigenous" or "native." Italy has a few other varieties that include the word "Vernaccia" in their name, but they are all unrelated.

FURTHER EXPLORATION

The San Gimignano wine-growing area overlaps the Colli Senesi sub-zone of the Chianti DOCG. Many local producers consequently make Vernaccia di San Gimignano as well as reds under the Chianti Colli Senesi DOCG.

THE PLACE

San Gimignano is one of Toscana's most distinctive and best-preserved medieval hilltop villages. It looks very much as it did in the 12th and 13th centuries and is among Toscana's most visited tourist attractions. The village lies at the center of the winegrowing area with vineyards covering the surrounding hills.

GRAPE VARIETIES

Vernaccia di San Gimignano is one of the oldest Italian grapes and has been highly regarded for centuries. The grape has historically been grown almost exclusively around the village of San Gimignano and this tradition continues today.

WINE PROFILE

What makes this wine special: The Vernaccia di San Gimignano DOCG is the embodiment of a symbiotic relationship between a single grape variety and a specific terroir

Acidity: Crisp acidity

Body: Medium- to full-bodied, medium alcohol

Common descriptors: Delicately aromatic with hints of citrus, white flowers, sage, ripe yellow fruit, almond, wet stones, and mineral; develops flint aromas with bottle age

Food pairings: Aperitif, seafood and shellfish, fritto misto (assorted fried items, usually fish and other seafood), first courses with white sauces and mushrooms, rice and pasta salads, white meat

PRODUCTION NOTES

- Wines must be crafted from at least 85% Vernaccia di San Gimignano

- The Riserva version must be aged for at least 11 months; most Riserva bottlings usually spend some time in oak and show a more complex and rounded profile with hints of vanilla and sweet spices

NOTABLE PRODUCERS

Sono Montenidoli: This producer makes wines in a historic style—rich, nutty and with alluring hints of oxidation; considered by many to be the best wines of the appellation

Fontaleoni: Produces bright and fresh wines that also have depth and texture; classic expressions of Vernaccia di San Gimignano

Guicciardini Strozzi: Run by the 15th generation descendants of Mona Lisa; the estate produces some of the best-known Vernaccia di San Gimignano on the wine market today

Panizzi: Their 1998 Riserva was the first Vernaccia di San Gimignano to be awarded Three Glasses by the distinguished Gambero Rosso Wine Guide

TOSCANA'S SIGNATURE CUISINE

BREAD

Pane Sciocco and Ribollita
Toscana is the only Italian region that makes a bread without any salt. Called "pane sciocco," this bread has been made since the Middle Ages and is an important ingredient in Toscana's famous "Ribollita." Ribollita starts out as a vegetable soup. Stale bread is added to any leftovers. As the concoction is reheated (or reboiled), the bread soaks up the broth and the soup turns into a thick, delicious stew.

SPICE

Saffron
San Gimignano is also well known for its precious saffron, the world's most expensive spice. Cultivated here since at least the 13th century, the spice was used as currency during the Middle Ages. In fact, some of the famous medieval towers of San Gimignano were built by families that made their fortunes in the saffron trade. The saffron of San Gimignano is considered among Italy's best.

ORVIETO DOC
Umbria & Lazio

Orvieto was once Umbria's most prestigious wine. Dubbed as the "sun of Italy in a bottle," it held a rightful place among Italy's classic white wines. Although its popularity waned with the march of time, its star is rising once again. The 21st century has ushered in a resurgence of interest in the appellation with the rise of a new generation of talented producers dedicated to restoring the ancient prestige of this remarkable terroir.

Orvieto makes up 80pc of Umbria wine - some interesting reds being developed - Montefalco Sagrantino (v. tannic)

THE PLACE

Orvieto is a striking town dramatically perched atop a cliff overlooking the surrounding valleys of southwestern Umbria. It is historically one of Italy's most important centers of art and culture. *principally in Umbria*

Umbria shares the Orvieto DOC with its southern neighbor Lazio, but Umbria claims most of the appellation's vineyards and makes most of the wine. The historic heart of the Orvieto DOC is the hilly and volcanic classico zone that surrounds the town.

GRAPE VARIETIES

two grape types

Orvieto is primarily crafted from Grechetto *(Greh-KET-toe)* and Trebbiano Toscano grapes. Grechetto is considered to be of higher quality and is the more distinctive of the two, but it is a little more complicated than that!

The name Grechetto is used to refer to two distinct grape varieties in Umbria – Grechetto di Orvieto and Grechetto di Todi. Historically, it was thought that the two were identical and thus the DOC rules make no distinction between them.

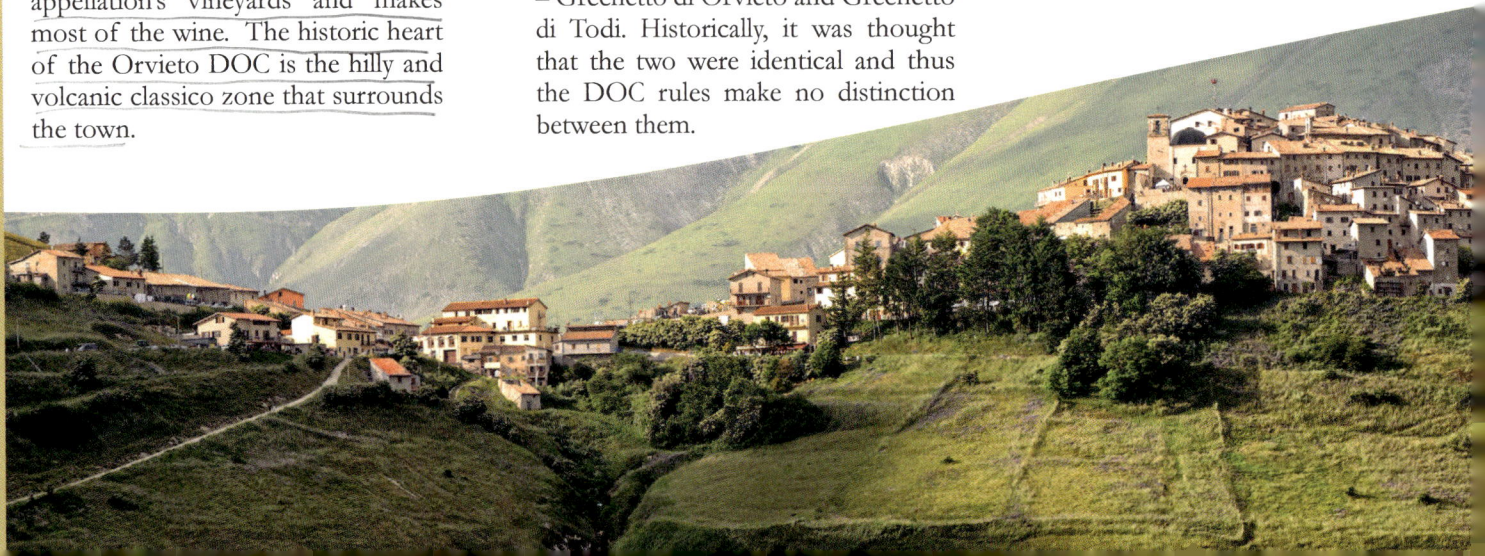

Producers often label their wines with the generic "Grechetto" and it can be impossible to know which grape has been used, or if the wine is a blend of both varieties. To make matters even more confusing, Grechetto di Todi is also known by the name Grechetto Gentile!

WINE PROFILE

What makes this wine special: The Orvieto DOC offers fresh, easy-drinking white wines. The best examples are floral with citrus and stone fruit, rich texture, depth and minerality.

Acidity: Lively, crisp acidity

Body: Light- to medium-bodied, medium alcohol

Common descriptors: Citrus (lemon, grapefruit), white flowers, apple, melon, peach, mineral

Food pairings: Appetizers, seafood, ham, pasta with cream sauce

NOTABLE PRODUCERS

Barberani: One of Umbria's most respected producers; famous for the pure flavors of their wines and for their botrytis-affected dessert wine, Calcaia

Palazzone: One of the region's best producers; this estate is known for its masterful blend of the five traditional varieties allowed in the Orvieto DOC

Decugnano dei Barbi: This estate's territory has been recognized for over eight centuries and linked to wine production since the 13th century. The soils are rich in fossilized oyster and seashells, which are said to contribute to the salty, mineral flavors of the estate's wines.

Sergio Mottura: One of Lazio's greatest wine producers with an instantly recognizable label featuring a porcupine—the winery mascot. Porcupines almost overran the estate when it converted to organic farming—a testament to the pristine environment that Mottura created. His Orvieto is often cited as the best white wine of Umbria and Lazio.

PRODUCTION NOTES

- Grechetto and Trebbiano Toscano must account for at least 60% of the blend; small proportions of other local grapes are permitted

- Orvieto can be designated as Superiore if stricter production regulations (such as higher minimum alcohol) have been met; many of the best examples of Orvieto are produced within this category

- Most Orvieto is unoaked

- Historically, Orvieto's most highly praised wines had a bit of residual sugar, but today, most wines are produced in a dry (secco) style

Pozzo di San Patrizio

UMBRIA'S SIGNATURE CUISINE

TRUFFLES

Umbria claims to be the single biggest producer of truffles in Italy. While white truffles are found here, the region is most famous for the black truffles that can be hunted nearly year-round. In fact, black truffles are so prevalent in Umbria that they are an integral part of the cuisine. While the rarer white truffles are saved for special occasions, black truffles are more of an everyday ingredient and their earthy flavor can be tasted in a plethora of Umbrian dishes and delights.

LENTILS

Umbria is also known for a special variety of lentil grown on the plains of Castelluccio and on the Colfiorito plateau. Similar to France's Puy lentils, they have a nutty flavor and hold their shape after cooking.

VERDICCHIO DEI CASTELLI DI JESI DOC
Marche

Verdicchio is one of Italy's noble white grape varieties. In Marche's Castelli di Jesi, the grape makes some of Italy's most noteworthy white wines.

ETYMOLOGY

In the Middle Ages, Jesi was a "Città Regia" (Royal City)—a city with special rights because it was the birthplace of a Holy Roman emperor. Jesi grew in power over time and eventually gained control of surrounding territories and numerous villages. These municipalities were characterized by hilltop medieval fortifications called "castelli" (castles) and the entire area became known as Castelli di Jesi (Castles of Jesi).

SHOPPING TIP

Wines labeled with just the Verdicchio dei Castelli di Jesi DOC name are generally light, quaffable and easy-drinking. For richer, more complex and longer-lived examples, look for the words "Classico Superiore" and/or "Riserva" on the label.

FURTHER EXPLORATION

Castelli di Jesi is often compared to the Verdicchio di Matelica DOC. The town of Matelica is also in Marche, but lies further inland, near the foothills of the Central Apennines. The winegrowing area lacks any temperature-moderating influence from the sea and growing conditions are cooler than in Jesi. Because of this, the wines show higher acidity and are more restrained and mineral in character than the broader styles of Castelli di Jesi. Verdicchio from Matelica is also thought to have long aging potential. The best and longest-lived wines are produced under the Verdicchio di Matelica Riserva DOCG which must be aged for at least 18 months.

THE PLACE

The Verdicchio dei Castelli di Jesi DOC encompasses an area close to the ancient town of Jesi. The appellation's vineyards lie on gentle hills and benefit from the climatic interplay between the Adriatic Sea 15 mi/25 km to the east and the cooling influence of the Central Apennines to the west.

There are 22 villages entitled to make wines under the DOC, but only vineyards in the historic and original winegrowing zone are entitled to the classico designation.

GRAPE VARIETIES

Verdicchio is Marche's flagship white variety and the region's most widely planted white grape. The name is believed to derive from the Italian word "verde" (green) and refers to the greenish color of the grapes.

What makes this wine special: Verdicchio is a versatile grape that is used to make a range of styles—from sparkling to sweet passito and everything in between. However, in Marche, the grape is at its best as a still, dry wine. The top examples have a remarkable combination of structure, richness, complexity and age-worthiness.

Acidity: Pronounced

Body: Medium- to full-bodied, medium to high alcohol

Common descriptors: Floral (chamomile), citrus, apple, melon, mineral and almond. With bottle age, the wine develops notes of almond paste, honey, beeswax and petrol

Food pairings: Appetizers, pasta, seafood, fish, white meat

PRODUCTION NOTES

- Verdicchio dei Castelli di Jesi DOC must contain a minimum of 85% Verdicchio; however, most are 100%

- The wines are mainly still and are usually made in a dry style, but sparkling (spumante) and sweet (passito) versions are also produced

- Verdicchio dei Castelli di Jesi DOC can be designated as Classico and Classico Superiore, if produced in the classico zone. Some of the best examples of Verdicchio are produced within the Classico Superiore category which must follow stricter production rules.

- Castelli di Jesi Verdicchio Riserva DOCG is the top quality tier of the Verdicchio dei Castelli di Jesi. It requires a minimum of 18 months of aging and often spends time in oak. The riserva can also be designated as Classico.

NOTABLE PRODUCERS

Umani Ronchi: Family-run estate that has succeeded in bringing Verdicchio to the wider world market

Villa Bucci: A benchmark producer; led by Renaissance man Ampelio Bucci

Tavignano: This organically certified estate's Misco 2015 Verdicchio dei Castelli di Jesi DOC Classico Superiore was awarded the Gambero Rosso's "Best Italian White Wine" in 2017

Andrea Felici: Winemaker Leopardo Felici crafts two wines, each with purity and pronounced minerality, and each showcasing the potential of Verdicchio in this unique terroir

La Staffa: Owner and enologist Riccardo Baldi crafts complex wines from organic vineyards

MARCHE'S SIGNATURE CUISINE

OLIVES

Olive all'Ascolana is a specialty of Marche. Olives are stuffed with meat and parmesan cheese, then breaded and fried. This delicious, salty-briny snack is the perfect accompaniment to Verdicchio dei Castelli di Jesi.

FISH

Brodetto
Fresh fish and seafood are an integral part of Marche's gastronomy. Brodetto is a notable regional fish soup that was originally made by fishermen, using the scraps of fish that could not be sold in the local fish markets. The tradition of brodetto has continued to evolve and now each coastal town on the Adriatic shore has its own variation. Arguably, the most famous is brodetto all'Anconetana. This version must contain 13 different types of fish and seafood which are cooked in a base of tomatoes, onions, garlic and herbs. Brodetto has become so important to Marche's culture that festivals are held throughout the year to celebrate its various incarnations.

Photo courtesy of Tavignano

Photos courtesy of Avignonesi

DETOUR

VIN SANTO DOCs
Toscana

Vin Santo is Toscana's classic dessert wine. This ancient passito traces its origins back to the ancient Greek and Roman practice of air-drying grapes in order to concentrate sugars.

THE PLACE

Vin Santo is produced across Toscana; however, the most exalted examples come from Carmignano, Chianti Classico, Chianti and Montepulciano. Each of these areas has its own appellation for the production of this wine:

- Vin Santo di Carmignano DOC
- Vin Santo del Chianti DOC
- Vin Santo del Chianti Classico DOC
- Vin Santo di Montepulciano DOC

GRAPE VARIETIES

Most Tuscan Vin Santo is made from the white grapes Trebbiano Toscano *(Treb-bee-AH-no Tohs-CAH-no)* and Malvasia *(Mal-vah-SEE-ah)*.

WINE PROFILE

What makes this wine special: Vin Santo stands among Italy's best and most distinctive dessert wines

What you need to know: While Vin Santo may be produced in various sweetness levels ranging from dry to lusciously sweet, most examples are in the sweet range. The wines are intense and velvety, with a pleasant oxidative character. They typically show complex aromas and flavors of nuts, honey, hay, dried fruit, licorice and spice.

Vinsantaia at Avignonesi

PRODUCTION NOTES

- Grapes for Vin Santo are dried (appassimento) on racks or straw mats in well-aerated rooms or by hanging them from rafters. The drying process can last for months and results in concentrated sugars and flavors.

- The very sweet must is placed in small, old barrels known as caratelli where a slow fermentation and long maturation take place; top examples of Vin Santo can rest in caratelli for a decade or longer

NOTABLE PRODUCERS

Vin Santo di Carmignano: Capezzana is an award-winning producer whose Vin Santo is a blend of Trebbiano and the very rare San Colombano variety

Vin Santo del Chianti Classico: Rocca di Montegrossi meticulously sort their grapes during the drying process to ensure that grapes are fully botrytized before crushing

Vin Santo del Chianti: Selvapiana makes a 100% Trebbiano Vin Santo that possesses an uncommon purity and freshness

Vin Santo di Montepulciano: Avignonesi produces exceptional Vin Santi that serve as benchmark examples; their Occhio di Pernice is a rare and ethereal wine

Appassitoio at Avignonesi

HOLY WINE

The name "Vin Santo" translates as "holy wine." Historically, the dried grapes were ready to be pressed and fermented by March or April—around the same time as the Christian Easter holidays. Eventually, it became traditional to do the wine work during the "holy week," and thus the wine became known as "holy wine."

FURTHER EXPLORATION

Occhio di Pernice, or "eye of the partridge," is a style of Vin Santo made from red grapes (predominantly Sangiovese). This rare wine is produced in very small quantities by only a handful of producers. Incredibly concentrated and complex, these exclusive wines are certainly worth seeking out. Avignonesi estimates that with the grapes used to produce one 375ml bottle of Occhio di Pernice Vin Santo, they could produce twenty-four 750ml bottles of red wine!

NOTES

CENTRAL ITALY
Red Wines

CENTRAL ITALY RED WINES

0 — 50 — 100 km
0 — 25 — 50 miles

N

EMILIA - ROMAGNA

SAN MARINO

Pesaro

Lucca

Firenze

Pisa

Chianti Classico DOCG

Bistecca alla Fiorentina

Urbino

MARCHE

Ancona

Cònero

Livorno

TOSCANA

Arno

Arezzo

Siena

Gubbio

Grotte di Frasassi

Cortona

Bolgheri DOC

Montepulciano

Perugia

Assisi

Montefalco Sagrantino DOCG

Ascoli Piceno

Lago Trasimeno

UMBRIA

Montalcino

Brunello di Montalcino DOCG

Val d'Orcia

Montefalco

Montepulciano d'Abruzzo DOC

Elba

Prosciutto Toscano

Orvieto

Norcia

Prosciutto di Norcia

Pescara

Gran Sasso

LAZIO

Lago di Bolsena

Black Truffles

L'Aquila

Rocca Calascio

ABRUZZO

Tarquinia

Lago di Bracciano

Tiber

Maccheroni alla Chitarra

Cerveteri

Tivoli

Arrosticini

Roma

Frascati

MOLISE

PUGLIA

TYRRHENIAN SEA

Ostia Antica

Campobasso

ADRIATIC SEA

CAMPANIA

Maremma

CENTRAL ITALY RED WINES:

THE ROAD MAP

① MONTEPULCIANO D'ABRUZZO DOC
Mon-teh-pool-CHAH-no Dah-BROOTS-soh

The region: ABRUZZO
Ah-BROOTS-soh

Although small in size and quite mountainous, Abruzzo is an important wine-producing region of central Italy. Here, the Central Apennines give way to two bands of lower-elevation hills that merge into a narrow coastline. Vineyards are found in the hills, interspersed with olive groves. Coastal areas benefit from the warm influence of the Adriatic Sea, but the warmth does not travel very far inland and the winegrowing areas are decidedly continental in climate.

Not to miss travel sites:
Pescara is Abruzzo's most populated city. Situated on the Adriatic, it is known for beach resorts and abundant seafood. The city's history precedes the Roman invasion, but its location made it vulnerable to attacks and it has been demolished and rebuilt several times over. Therefore, it is one of the most modern cities (architecturally speaking) on the Adriatic coast.

Abruzzo has an abundance of natural beauty and much has been preserved by Italy's national park system. Gran Sasso and Monti della Laga National Park is one of Abruzzo's most popular destinations. Corno Grande (the Big Horn) is the highest peak of both the Gran Sasso massif and the Apennines, reaching more than 9,500 ft/2,900 m above sea level.

Castello di Roccascalegna is a 16th century medieval castle built right into the basalt rock of a mountain spur. Most likely built to protect the valley it lords over, the castle is a marvel of medieval engineering.

② CHIANTI CLASSICO DOCG
KYAHN-tee KLAHS-see-koh

③ BRUNELLO DI MONTALCINO DOCG
Broo-NEHL-lo Dee Mon-tahl-CHEE-noh

④ BOLGHERI DOC
Bol-GEH-ree

The region: TOSCANA/TUSCANY
Tohs-CAH-nah

The Italian peninsula truly begins in Toscana, and the region displays a vast mosaic of distinctive terroirs. The Northern Apennines curve around northern Toscana then begin the journey south as Italy's "spine." The easternmost reaches of the region are quite mountainous, but hills cover the majority of the Toscana landscape. As in Piemonte, the major groups of hills have names and are home to the majority of the region's vineyards, including the Chianti Classico and Brunello DOCGs. However, the hills soften and fade into coastal plains which famously host some of Toscana's most famous vines—those of Bolgheri and Sassicaia.

Not to miss travel sites:
The town of Siena lies in the center of the Chianti region and is renowned for its principal public square, Piazza del Campo, which is regarded as one of Europe's greatest medieval gathering spaces.

Monte Amiata towers over southern Toscana and is among Italy's highest volcanos. Long dormant, the peak and its surrounding area are loaded with hiking paths, ski trails and thermal spas, making it a popular destination for outdoors enthusiasts. The Amiata area is also dotted with medieval villages, known for showcasing the local gastronomy.

⑤ MONTEFALCO SAGRANTINO DOCG
Mon-teh-FAHL-koh Sah-GRAHN-tee-noh

The region: UMBRIA
OOM-bree-ah

The Central Apennines run along Umbria's northern and eastern borders, blocking both cool northern air and warmer Adriatic Sea breezes. However, the lower-elevation Sub-Apennines range in the west allows temperature-moderating warmth from the Mediterranean to reach into the central and southern parts of the region where most of the vineyards are found.

Not to miss travel site:
Perugia is Umbria's capital and this ancient city is a cultural and artistic treasure known for its museums, music festivals, and Etruscan and medieval architecture. The University of Perugia is one of the oldest in the world. However, in some circles, the city is perhaps best known for the production of chocolate.

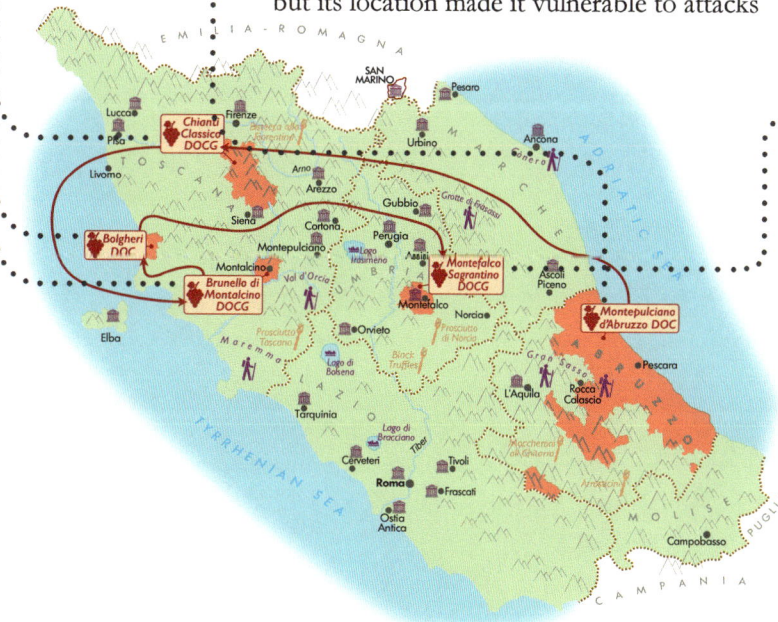

MONTEPULCIANO D'ABRUZZO DOC
Abruzzo

The Montepulciano d'Abruzzo DOC is one of Italy's largest by volume and is mainly known for fruity, easy-going reds. However, this appellation also makes serious, powerful and complex bottlings that include some of the most coveted wines produced in Italy.

THE PLACE

The Montepulciano d'Abruzzo DOC is a regional appellation that covers Abruzzo's entire winegrowing area. Vineyards lie on hills that extend inward from the Adriatic coast to the foothills of the Apennines. The growing area benefits from the interplay between the Adriatic Sea to the east and the cooler air masses of the mountain peaks to the west. As a result, diurnal temperature swings are significant and there are constant breezes blowing throughout the region. These ideal conditions yield healthy, fully ripe grapes that have retained natural acidity and developed aromatic complexity.

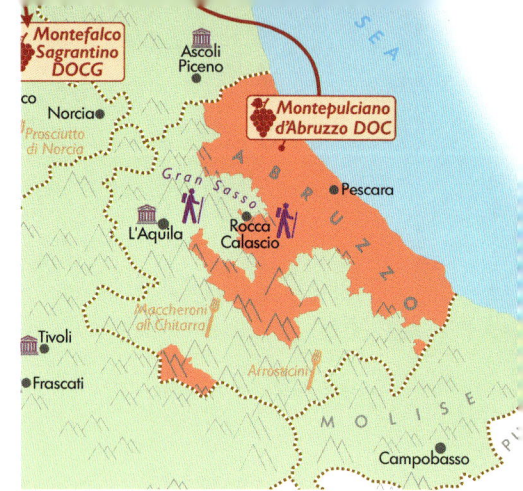

GRAPE VARIETIES

Montepulciano is native to Abruzzo. It is the region's claim to fame and one of Italy's most widely planted grapes.

FURTHER EXPLORATION

The Cerasuolo d'Abruzzo DOC produces one of Italy's best rosé (rosato) wines. It is made with Montepulciano, a grape that is ideal for rosé production because of the high level of pigment in its skin and the nearly pink color of its juice!

THE APENNINES

The Apennines are the principal mountain ranges of Central and Southern Italy. They run through the center of the Italian peninsula, forming its backbone. They are lower in altitude than the Alps; however, many peaks are 6,500 ft/2,000 m above sea level.

The Apennines are generally divided into three contiguous groups: the Northern Apennines, which begin in Liguria and run through Lombardia, Emilia-Romagna, and a portion of Toscana, the Central Apennines which trek through Marche, Umbria, and Abruzzo, and the Southern Apennines, which trail through the remainder of the peninsula terminating on the island of Sicilia.

WINE PROFILE

What makes this wine special: Montepulciano d'Abruzzo wines are produced in a wide range of styles from light, easy-drinking and fruit-forward to significantly more robust, fleshy and oak-aged

Acidity: Marked acidity

Tannin: Medium to high tannins

Body: Medium- to full-bodied, medium to high in alcohol

Common descriptors: Deeply pigmented, cherry, plum, savory, earthy, leather, tobacco, dark chocolate

Food pairings: Pizza, pasta with meat sauce, lamb and mutton, braised meat. Structured, full-bodied versions pair well with richer, more complex dishes.

PRODUCTION NOTES

MIN 85 most 100pc

- A minimum of 85% Montepulciano is required, although most examples are 100%

- Montepulciano d'Abruzzo can be produced as Riserva with a minimum of two years of aging of which nine months are spent in wood

NOTABLE PRODUCERS

Valentini: Iconic producer whose Montepulciano is widely considered one of Italy's greatest wines

Emidio Pepe: Historic, benchmark producer and a pioneer of biodynamics within Italian viticulture

Tiberio: Pioneering producer that is investing in the region and producing fresh, focused examples of Montepulciano d'Abruzzo

Illuminati: Family-owned winery that has embraced innovation throughout its 120-year history

ABRUZZO'S SIGNATURE CUISINE

PASTA

Maccheroni alla chitarra is an egg noodle named for the tool used to make it—the "chitarra" which resembles a guitar. It is typically served with meat ragù.

Three of Italy's best-known brands of pasta (De Cecco, Delverde and Cocco) are based in Abruzzo. The picturesque village of Fara San Martino has a long-standing reputation for making exceptional pasta.

MEAT

Arrosticini are skewers of meat (typically mutton) that are grilled over a charcoal flame. These tasty morsels are eaten straight off the skewer and are considered one of central Italy's most famous street foods. Typical of mountain fare, both maccheroni alla chitarra and arrosticini are linked to Abruzzo's pastoral traditions.

The wines of the Chianti Classico DOCG and the Chianti DOCG are often mistakenly considered as one and the same.

• The Chianti Classico DOCG is the original and historic production area for wines that have been known as Chianti since the Middle Ages.
• Wines labeled as Chianti DOCG hail from land surrounding the original zone of production. During the 1930s, it was decided to expand the original Chianti production zone. The original zone was given the name "Chianti Classico" and the newly expanded areas were given the right to call their wines "Chianti."

FURTHER EXPLORATION

Chianti DOCG is produced in significantly larger volumes than Chianti Classico DOCG. As a general rule, the wines are inexpensive, fruity and light, and meant for early consumption. They require a minimum of 70% Sangiovese. The appellation is a large area that loosely surrounds the smaller, hillier Chianti Classico DOCG and includes seven official sub-zones. Among the most well known are Rùfina, Colli Senesi and Colli Fiorentini. Chianti Rùfina, in particular, is a historic winegrowing area that makes refined and age-worthy wines that can rival the wines of Chianti Classico.

Central Italy Red Wines

CHIANTI CLASSICO DOCG
Toscana

Chianti Classico is one of Italy's most historic and famous wines. It is also one of the world's oldest demarcated wine appellations—it was originally delimited in 1716.

THE PLACE

The winegrowing area of the **Chianti Classico DOCG** lies in the hilly, central part of Toscana, between the cities of Firenze (Florence) and Siena and stretches for more than 40 mi/65 km from north to south.

The landscape in this part of Toscana features an iconic mix of hills and valleys covered with vines, olive groves, pine forests, chestnut trees, and stately rows of cypress. Medieval castles dot the hilltops and provide spectacular views of the romantic countryside.

GRAPE VARIETIES

The wines of the Chianti Classico DOCG are based on Sangiovese *(Sahn-jo-VAE-sae)*. It is Toscana's signature grape and is by far Italy's most widely planted grape variety.

What makes this wine special: The top bottlings of the Chianti Classico DOCG proudly stand among the most admired red wines of Italy. They possess a tangy acidity that makes them incredibly versatile food wines.

Acidity: High

Tannin: Medium to high

Body: Medium-bodied, medium alcohol

Common descriptors: Violet, sour cherry, plum, tea leaves, licorice and earth. The wines develop notes of leather and tobacco with bottle age.

Food pairings: Pasta with tomato sauce, barbecue, hamburgers, steaks

PRODUCTION NOTES

MIN 80 PC

- Chianti Classico must include a minimum of 80% Sangiovese. A small proportion of local grape varieties such as Canaiolo or international grapes such as Merlot or Cabernet Sauvignon can be added as minor blending partners. (Some producers opt to craft Chianti Classico that is 100% Sangiovese.)

- Chianti Classico is produced in three versions that represent three tiers in the appellation's quality pyramid:

  - **Chianti Classico:** Younger wines made in a lighter style
  - **Chianti Classico Riserva:** This wine must be aged for at least 24 months and is fuller and more complex with longer aging potential
  - **Chianti Classico Gran Selezione:** Grapes used must be exclusively estate grown and the wine must be aged for a minimum of 30 months; it is meant to represent the best wines within the Chianti Classico appellation

NOTABLE PRODUCERS

Montevertine: Iconic producer responsible for creating Le Pergole Torte in 1977; this 100% Sangiovese wine was revolutionary as Chianti Classico was required to contain a percentage of white grapes at that time

Ricasoli: Historic estate run by the Ricasoli family, whose ancestor, Baron Bettino Ricasoli, was one of Toscana's original wine entrepreneurs and the man responsible for developing the first blending formula for Chianti

Monteraponi: Michele Braganti crafts elegant and nuanced wines that are Burgundian in their approach at a high-elevation estate just outside the village of Radda

Castellare: An estate just outside the town of Castellina; dedicated to environmentally sustainable cultivation and traditionally made wines

Fèlsina: Award-winning producer famous for two monovarietal Sangiovese wines, Fontalloro and Rancia

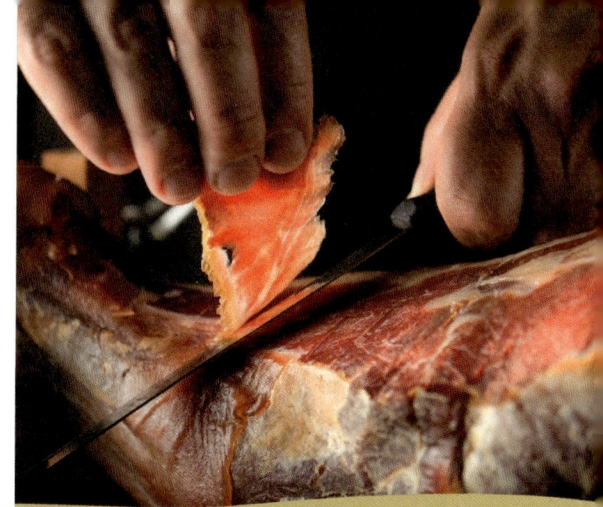

CHIANTI CLASSICO
GRAN SELEZIONE

CHIANTI CLASSICO
RISERVA

CHIANTI CLASSICO

TOSCANA'S SIGNATURE CUISINE

BREAD

Pappa al Pomodoro
Stale bread is transformed into several signature Tuscan dishes. This popular appetizer consists of stale bread, peeled tomatoes, garlic, basil and olive oil. Although the ingredients are humble, the combination creates something wholesome and satisfying.

MEAT

Prosciutto Toscano DOP
Toscana has a long-standing tradition of dry-curing ham and rules regarding production have been in place since the 15th century! The delicious Prosciutto Toscano DOP is made across the region and is generally described as slightly saltier than other prosciuttos. However, its saltiness makes it a perfect match with Toscana's salt-free bread.

BRUNELLO DI MONTALCINO DOCG
Toscana

Brunello di Montalcino stands among Italy's most prestigious and famous wines. It is considered the most structured, powerful and longest-lived expression of the Sangiovese grape.

highest bit of Tuscany Sangiovese achieve maximum ripeness here

THE PLACE

Montalcino is a medieval hill village in southeastern Toscana (about 25 mi/40 km south of Siena). The town overlooks the surrounding hills and valleys offering a charming view of the typical Tuscan landscape.

Vineyards are planted at varying elevations over a series of irregular hills, slopes and ridges surrounding the village of Montalcino. The combination of warm, dry days and cooler nights allows Sangiovese to ripen fully while preserving the grape's natural high acidity.

GRAPE VARIETIES

Brunello di Montalcino is made from Sangiovese, locally known as Brunello. The name Brunello, meaning "little brown one," refers to the dusky brown color of the fully ripened berries.

ETYMOLOGY

The Sangiovese grape has many synonyms in Toscana—the name changes depending on where the grape is grown. Common names include:

- Brunello (in Montalcino)
- Prugnolo Gentile (in Montepulciano)
- Morellino (in Scansano)

FURTHER EXPLORATION

Rosso di Montalcino DOC is essentially a "second wine" to that of Brunello di Montalcino - much like the second labels for the "grand vin" in Bordeaux. The zone of production corresponds to that of Brunello. Rosso di Montalcino is also made from 100% Sangiovese but the grapes are usually sourced from lesser or younger vineyards. It can also be made with wines that do not meet the qualitative standards deemed necessary for Brunello. Rosso di Montalcino DOC is generally a lighter, fruitier and more approachable version of its "older brother" and is less expensive.

SHOPPING TIP

In poor vintages, when producers opt not to make Brunello, Rosso di Montalcino may be a good bargain because Brunello grapes are often declassified and incorporated into that year's Rosso di Montalcino.

Photo courtesy of Podere Le Ripi

WINE PROFILE

What makes this wine special: Nowhere else is Sangiovese able to achieve the same combination of finesse, depth, power, complexity and aging potential as in Montalcino

Acidity: High acidity

Tannin: High tannins

Body: Full-bodied, high alcohol

Common descriptors: Floral (violet), sour red cherry, plum, licorice, dried flowers, underbrush, balsam, tobacco and leather

Food pairings: Steaks, game, aged cheese

PRODUCTION NOTES

- 100% Sangiovese
- Brunello di Montalcino has the longest aging requirements in Italy; the wines must age for at least four years, two years of which must be in oak
- Riserva versions are aged for an additional year.

NOTABLE PRODUCERS

Biondi-Santi: The history of Brunello di Montalcino is deeply connected to the Biondi-Santi family, founders of one of Italy's greatest wine estates. The family is credited with giving Montalcino a spotlight on the world wine stage when they bottled the first wine labeled Brunello in the late 1800s.

Il Marroneto: Il Marroneto makes pure, polished and refined Brunello wines that are true to Sangiovese's character with refreshing acidity and a dusty mineral core. These are some of the most traditionally styled Brunellos being made today.

Le Ragnaie: This young, high-elevation estate owned by Riccardo and Jennifer Campinoti uses Burgundian winemaking techniques to create delicate, silky wines. A winery to watch!

Il Poggione: This estate is one of the three original producers of Brunello di Montalcino and its origins date back to the late 1800s. It is known for rich, ripe and structured wines.

Case Basse Soldera: An iconic producer of wines from 100% Sangiovese produced outside of the Brunello di Montalcino DOCG. The wines have developed cult status for their concentration, richness, elegance and complexity.

Poggio di Sotto: Located in the warmer Castelnuovo dell'Abate part of Montalcino, this estate crafts expansive wines that beautifully balance power and precision

Podere Le Ripi: Biodynamic winery famous for its Bonsai vineyard (62,500 vines/ha) and its Golden Cellar (constructed with ancient building techniques)

TOSCANA'S SIGNATURE CUISINE

MEAT

Bistecca alla Fiorentina is one of Toscana's most famous food specialties. The beef comes from a renowned breed of white Tuscan cattle known as Chianina. Grilled over hot coals, the massive steak is minimally seasoned with local herbs and salt (or with no seasonings at all, depending on the chef). Experts recommend bistecca alla fiorentina be served rare, in order to preserve the tenderness of the meat. This Tuscan treat is classically paired with Sangiovese-based wines such as Brunello di Montalcino or Chianti Classico.

BOLGHERI DOC
Toscana

Bolgheri is one of Italy's most renowned areas for fine red wines. It is the origin of the Super Tuscan movement and produces some of Italy's most iconic wines.

THE PLACE

Bolgheri is a tiny, picturesque hamlet in the northern coastal portion of the Maremma. The appellation lies along a strip of largely flat coastal land. The vineyards enjoy plenty of sunshine and the sea provides cooling breezes, which is especially important during the warmth of the growing season.

These conditions allow the grapes to ripen fully yet maintain enough acidity to craft a wine with freshness and vitality.

GRAPE VARIETIES

The wines of the Bolgheri DOC are generally Cabernet Sauvignon, Merlot and Cabernet Franc blends. Small proportions of Sangiovese, Petit Verdot and Syrah are occasionally included.

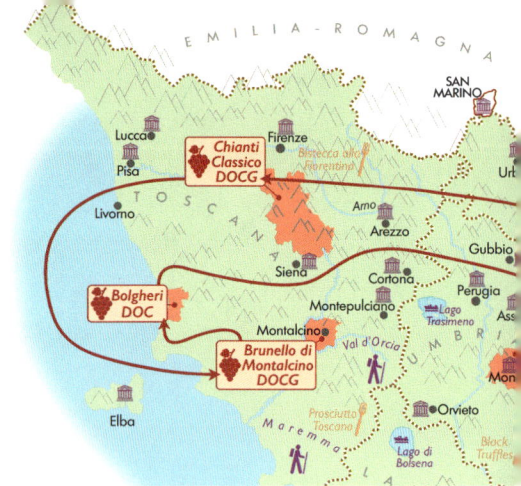

WINE PROFILE

What makes this wine special: The wines of the Bolgheri DOC are Bordeaux-style blends that offer a combination of lush, concentrated fruit and sweet-oak notes coupled with a distinctive hint of Mediterranean garrigue (resinous herbs)

Acidity: Medium

Tannin: Dense, velvety tannins

Body: Rich, full-bodied and high in alcohol

Common descriptors: Ripe black fruit, balsamic notes, baking spices, hints of Mediterranean scrub

Food pairings: Charcuterie, red meat, game and aged cheeses

PRODUCTION NOTES

- The Bolgheri DOC blending rules are flexible—producers can opt for any proportion of Cabernet Sauvignon, Merlot or Cabernet Franc. However, Sangiovese and Syrah can only account for a maximum of 50% of the blend.

- Wines made under the Bolgheri Rosso DOC must be aged for one year, while those carrying the Bolgheri Rosso Superiore DOC label must be aged for a minimum of two years (with at least one of those in oak). The superiore version is generally an estate's top wine.

- A small amount of white Bolgheri DOC is made; this wine is typically a Vermentino-based blend

25 DOC
BOLGHERI
CONSORZIO DI TUTELA

NOTABLE PRODUCERS

Tenuta dell'Ornellaia: Ornellaia truly is one of the world's iconic wineries; it is owned by the Frescobaldi family

Fabio Motta: This up-and-coming producer is crafting elegant yet powerful wines. Motta has worked for years alongside legendary winemaker, Michele Satta. One to watch.

Le Macchiole: This boutique winery, led by Cinzia Merli, is best known for single-varietal wines. The most famous of these is Paleo, a monovarietal Cabernet Franc that is considered one of the world's best expressions of the grape.

Michele Satta: Choosing to focus on Sangiovese and Syrah, the wines of Michele Satta offer a unique take on the terroir of Bolgheri; the winery is now in the hands of the second generation

Grattamacco: This estate is one of the appellation's founding wineries; it is currently undergoing a period of revitalization after being sold to new owners

Podere Castellaccio: Organically farmed estate that has some of the oldest vines in Bolgheri, including the less commonly seen Ciliegiolo, Foglia Tonda and Pugnitello

TOSCANA'S SIGNATURE CUISINE

🥩 MEAT

Cinghiale
Wild boars (cinghiale) roam freely in Toscana, especially in the Maremma. Boar hunting is a favored pastime (the hungry beasts do a lot of damage to local agriculture) and the meat is a staple of the local cuisine. Many towns and villages hold annual festivals to celebrate this animal and there are countless ways to cook it. Structured Super Tuscans are a perfect pairing with stews and braised or grilled boar meat.

🍝 PASTA

Pappardelle con ragù di cinghiale alla toscana (large, broad ribbons of pasta with a ragù of wild boar) ranks at the very top of Toscana's signature boar dishes. Wild boar leg or shoulder is marinated in red wine to tenderize and flavor it, then braised with crushed tomatoes, garlic, onion, celery, rosemary, and juniper.

MONTEFALCO SAGRANTINO DOCG
Umbria

Montefalco Sagrantino is Umbria's flagship red wine and stands among the most distinctive and unique red wines of central Italy.

ETYMOLOGY

The origin of the name "Sagrantino" is believed to derive from the words "sacer/sacro" (sacred) which implies that the wine was used for religious ceremonies and celebrations.

BEYOND WINE

Sagrantino International Challenge Cup

Each summer, the rolling hills of Umbria play host to the largest hot-air balloon gathering in Italy. It is a race to the sky and a race to the finish line! But there is competition on the ground also! A photography contest is held in conjunction with the aerial display. But that's not all! The event is sponsored, in part, by the local wineries. The vinous delights of Umbria are also showcased. Participants and spectators come from all over the world.

THE PLACE

Montefalco is a picturesque, medieval village in the foothills of the Central Apennines near the center of Umbria. The village is often called "La Ringhiera dell'Umbria" (Umbria's balcony) thanks to its panoramic, hilltop position which offers a breathtaking view of the surrounding valleys and distant mountain ranges. The vineyards of the appellation are interspersed with olive groves and flank the hills around Montefalco and four neighboring villages.

GRAPE VARIETIES

Sagrantino is an ancient red variety native to Umbria and known for its high level of tannin. Historically, the grape was mainly used to make limited quantities of sweet passito wine and was traditionally reserved for special occasions. Over time, this wine style fell out of favor.

The grape had almost disappeared by the 1960s, but was resuscitated by producers who started making dry wines with it. By the 1990s, dry Sagrantino achieved worldwide acclaim.

WINE PROFILE

What makes this wine special: Sagrantino is grown almost exclusively around the area of Montefalco, where it delivers intensely structured, powerful and long-lived wines

Acidity: Lively acidity

Tannin: Very high tannins

Body: Full-bodied and high alcohol

Common descriptors: Deep ruby color, black fruit (cherry, plum, blackberry), leather and licorice, with earthy, meaty, spicy notes

Food pairings: Roasted meat, stew, game, aged cheeses

NOTABLE PRODUCERS

Arnaldo Caprai: The Caprai estate is one of the historic leaders of the appellation and produces top-quality Sagrantino

Tabarrini: This was the first estate to bottle single-vineyard Sagrantino in Umbria, and it continues to produce some of the region's best wines

Antonelli San Marco: One of the oldest estates in the appellation, Antonelli San Marco is known for a slightly softer and refined style of Sagrantino

PRODUCTION NOTES

100%

- Most Montefalco Sagrantino is made as a dry wine; however, a tiny quantity of the traditional sweet version (passito) is still produced

- Both dry and sweet versions must be 100% Sagrantino

- Montefalco Sagrantino must be aged for a minimum of 37 months, including 12 months in oak. Oak maturation helps to soften the strong tannic structure of the wine.

- The grapes for Montefalco Sagrantino Passito must be air-dried for at least two months. The wine must be aged for a minimum of 37 months, but no oak aging is required.

UMBRIA'S SIGNATURE CUISINE

Umbria is one of Italy's most important sources of black truffles and the village of Norcia offers some of the region's best. Considered Umbria's culinary capital, the village also has a storied reputation for charcuterie. The local porchetta (pork roast) and the famous Prosciutto di Norcia IGP (dry-cured ham) are well-known delicacies.

PASTA

Stringozzi and **ciriole** are two of Umbria's typical pastas. Originally made without eggs and salt, these pastas truly reflect their peasant origins. (Eggs and salt were pantry ingredients of wealthy households!) Both are made from a simple combination of soft wheat flour (and sometimes durum wheat flour) and water.

While the techniques used to shape the pastas vary, ciriole is traditionally rolled flat (with a rolling pin) and then cut into thin, almost square strips. Stringozzi is even more irregular in shape as individual pieces are rolled out by hand and result in something resembling chubby spaghetti. However, the etymology of their names is even more interesting than their shape! Ciriole is apparently derived from the ancient name for a small, thin, white eel that the pasta resembles. Stringozzi, also known by the name strozzapreti, or "priest strangler," was apparently named in honor of the shoelaces that were used to strangle despised priests in the days of Pontifical rebellion.

NOTES

SOUTHERN ITALY
White Wines

SARDEGNA

Olbia

Alghero

Nuoro

Punta La
Marmora
6,017 ft/
1,834 m

GENNARGENTU

Cagliari

LAZIO

MOLISE

APENNINES

Foggia

Benevento

PUGLIA

Bari

Caserta

Taurasi

Avellino

Napoli

Monte Vesuvio
4202 ft/1,281 m

Alberobello

Brindisi

Ischia

Sorrento

Salerno

Potenza

Matera

Capri

CAMPANIA

BASILICATA

Taranto

ADRIATIC SEA

TYRRHENIAN SEA

Serra Dolcedorme
7,438 ft/2,267 m

IONIAN SEA

CALABRIA

CALBRIAN APENNINES

Catanzaro

0   30   60   90 km

0   25   50 miles

N

Messina

Palermo

NEBRODI

Taormina

Marsala

Monte Etna
10909 ft/
3,326 m

SICILIA

Catania

Agrigento

Monte Iblei
10909 ft/
3,326 m

Siracusa

Avola

Ragusa

Pantelleria

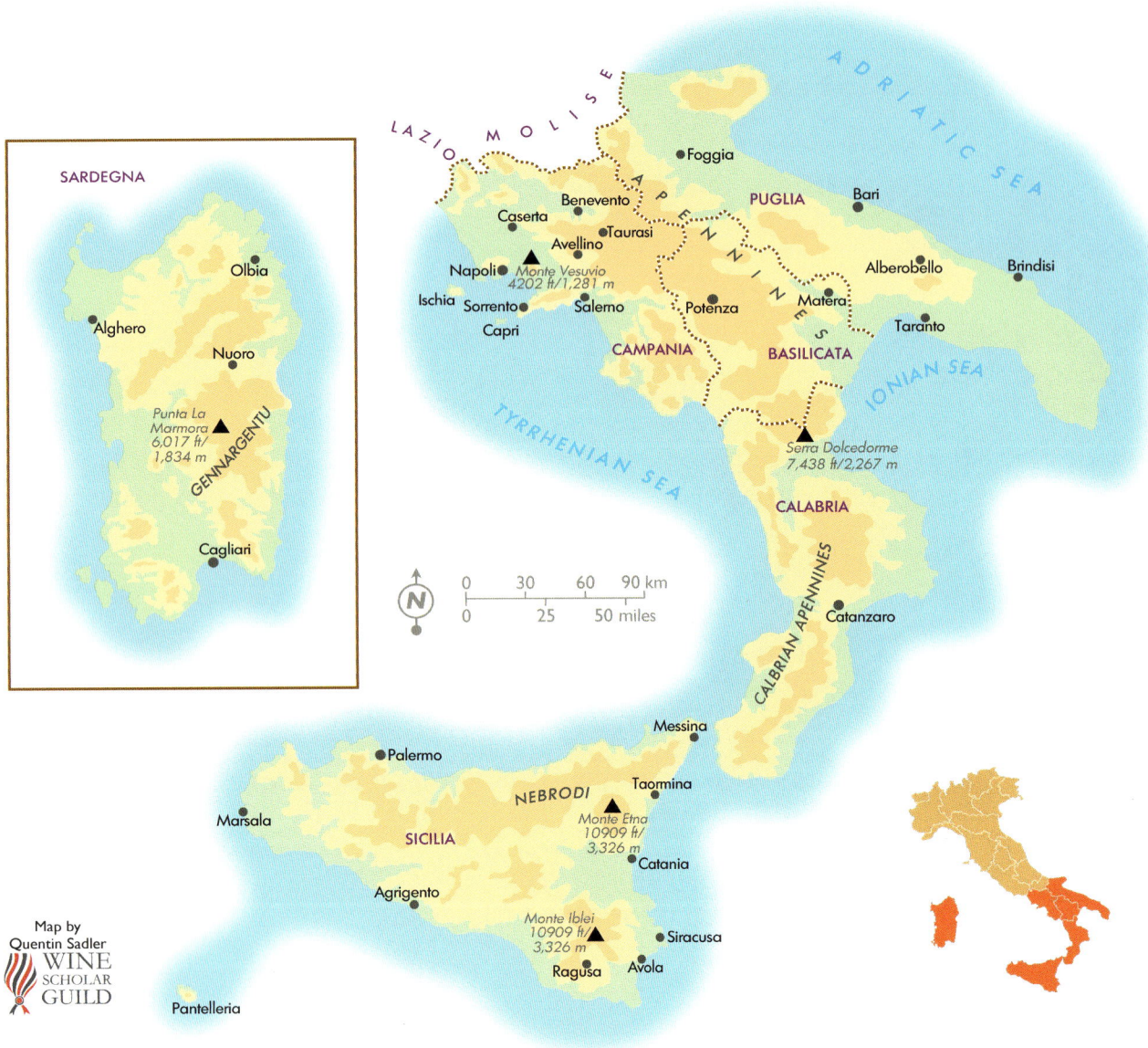

Map by
Quentin Sadler
WINE
SCHOLAR
GUILD

SOUTHERN ITALY

"If you like Italy as far south as Rome, go further south. It gets better. If Italy is getting on your nerves by the time you get to Rome, think twice about going further. Italy intensifies as you plunge deeper."
Rick Steves, Travel Writer, TV Producer

Greeks, Romans, Arabs, Normans, Spanish, Austrians, and many others have all contributed to the colorful culture of southern Italy. The language, architecture, arts and cuisine have all been enriched by this kaleidoscope. There seems to be more of almost everything in the south. As Rick Steves asserts, Italy gets more intense the further south you travel. The flavors of the food deepen, the sun becomes hotter, the traffic more chaotic, dialects more pronounced; there is a raw authenticity to the south that the highly polished north does not possess.

The topography of Italy's south is perhaps best described as rugged. The Southern Apennines continue their path along the western coastline and into Sicilia, creating a mountainous landscape that is often affected by seismic activity. Narrow coastal plains hug the water's edge while Campania's Monte Vesuvio and Sicilia's Monte Etna scratch the sky. Rivers are few and far between.

In stark contrast to the industrialized north, agriculture continues to be a mainstay of the south. This is largely due to the Bourbons who ruled the Kingdom of the Two Sicilies at the time of Italy's unification. The Bourbons were determined to keep their subjects isolated from the agricultural and industrial revolutions that were sweeping the north in order to preserve the existing feudal system. This dichotomy between north and south created a gap in both wealth and worldliness.

While transportation networks developed to support the burgeoning industrialization of the north, little development occurred in the south. As a result, even today, getting around southern Italy can be quite challenging. Yet, the south is a tourist mecca. It is home to some of the world's best-preserved Greek temples and Roman ruins, pristine, sandy beaches, volcanoes and turquoise blue waters.

The history of the south is mirrored in its cuisine. Considered to be the poorest part of Italy, much of the population has traditionally subsisted on largely vegetarian diets. Conversely, the nobility, with their enormous wealth, enjoyed a much more lavish culinary experience. Some of the most universally well-known Italian foods such as pizza, eggplant parmesan, focaccia and cannoli all originate in the south, and while substantial regional differences exist, there are some common denominators.

The southern latitude enables the ripening of heat-loving crops such as tomatoes and eggplants. These key ingredients are often found incorporated into the area's many pasta sauces. Pasta, in the south, is often cooked from a dried state, as opposed to the fresh egg-based pastas of areas further north.

The grass-covered hills of southern Italy have hosted flocks of sheep and herds of goats for centuries, so lamb and goat meat are popular additions to the southern table, as is the delicious ricotta cheese, which can be made from the milk of both. As one would expect, coastal areas rely heavily on the bounty of the sea; octopus, sardines and swordfish are mainstays. Olive oil plays a central role in the cuisine of the south and olive groves are a common sight throughout the region. Citrus fruits are commonly used to flavor dishes and are used to craft the region's famous limoncello. Beans, cured meats, hard cheeses and vegetables preserved in olive oil are all typical and say much about the region's difficult past. Peasants needed to preserve food in order to guard against a poor harvest.

Surprisingly, this simple cuisine has a rich dessert tradition. Desserts are popular and are generally much more extravagant than in the north! Sugar—introduced by the Arabs—arrived in this part of Italy first and was enthusiastically embraced!

Italy's south loves to eat. It lives to eat—but food here is much more than sustenance; it is considered an expression of love.

SOUTHERN ITALY WHITE WINES

0   30   60   90 km
0   25   50 miles

SARDEGNA

Zuppa Gallurese
Costa Smeralda
Vermentino di Gallura DOCG
Olbia
Grotta di Nettuno
Alghero
Nuoro
Culurgiones
Su Nuraxi
Porcetto Arrosto
Cagliari

LAZIO **MOLISE**

Burrata
Foggia
PUGLIA
Bari
Caserta
Greco di Tufo DOCG
Spaghetti alle Vongole
Taurasi
Avellino
Castel del Monte
Grotte di Castellana
Alberobello
Brindisi
Napoli
Vesuvio
Fiano di Avellino DOCG
Potenza
Ischia
Sorrento
Salerno
Matera
Capri
Insalata Caprese
Costiera Amalfitana
Taranto
Olive Oil
CAMPANIA
Mozzarella di Bufala
BASILICATA

ADRIATIC SEA

IONIAN SEA

TYRRHENIAN SEA

Catanzaro

CALABRIA

Messina

Marsala DOC
Palermo
Cannoli
Taormina
SICILIA
Etna
Marsala
Catania
Pasta con le Sarde
Agrigento
La Valle dei Templi
Siracusa
Avola

Baci di Pantelleria
Passito di Pantelleria DOC
Pantelleria

SOUTHERN ITALY WHITE WINES:

THE ROAD MAP

2 **FIANO DI AVELLINO DOCG**
FYAH-noh dee Ah-vehl-LEE-noh

3 **GRECO DEL TUFO DOCG**
GREH-koh dehl TOO-fo

1 **VERMENTINO DI GALLURA DOCG**
Vehr-mehn-TEE-noh dee Gahl-LOO-rah

The region: SARDEGNA/SARDINIA
Sahr-DEH-nyah

The island region of Sardegna offers one of the most beautiful and untouched landscapes in Italy. Large swaths of the island remain uninhabited and are characterized by coastal cliffs, weathered mountains of granite rock, and inland hills and forests. Yet, some of Europe's most beloved and pristine beaches are found in the northern part of the island along the "Costa Smeralda" (the Emerald Coast), a popular summer destination. France's mistral wind blows steadily across Sardegna, moderating seasonal heat and reducing humidity—appreciated by tourists and grapevines alike.

Not to miss travel sites:

The stunning beauty of Costa Smeralda has made it one of Europe's most popular and most expensive destinations. Private clubs, luxury hotels and amenities such as helicopter landing pads have attracted the "jet set," who are happy to pay a premium for exclusive access to pristine white-sand beaches and crystalline blue water.

La Maddalena is the largest and only inhabited island within a complex of 60 islands and islets that make up an archipelago off the northern coast of Sardegna. The island is known for its plentitude of small bays, coves and inlets, in addition to its beautiful beaches such as Cala Spalmatore and Cala Francese.

The region: CAMPANIA
Kahm-PAH-nee-ah

Campania is part of a volcanic arc that curves through southern Italy, and indeed, its landscape was shaped by volcanic activity. This is, after all, the region of Vesuvio, the volcano that buried the town of Pompeii in 79 C.E. Eruptions changed the composition of the soil surrounding the volcano, while wind carried ash to remote corners. The volcanic influence is felt just about everywhere in Campania—from the coastal plains and cliffs to the foothills of the Southern Apennines.

Not to miss travel sites:

One of Italy's most visited sites, the ancient city of Pompeii is also one the world's greatest archeological treasures. Buried in ash and pumice by the volcanic eruption of Monte Vesuvio, the partially excavated city offers today's world a rare glimpse into ancient life. Visitors not only explore public places such as an amphitheater, bathhouses and a central grassy forum, but also the preserved private homes and shops that line the narrow streets. The priceless information that has been uncovered during excavations has given historians incomparable insight into the evolution of every facet of this early society, including religion, economics and architecture.

Monte Vesuvio itself also offers an incredible opportunity for vulcanologists as one of the world's oldest and most-studied active volcanoes. For those more interested in recreation, a trip to the mountain's crater offers a stunning view of Naples, the Campanian countryside and the sea.

The Amalfi coast is rightfully considered one of Italy's most dramatic vistas. Its precipitous cliffs with terraced vineyards and lemon orchards is world-famous. Designated as a UNESCO World Heritage Site, this strip of breathtaking coastline includes 13 villages built on impossibly steep cliffs that seemingly rise straight from the astonishingly blue sea.

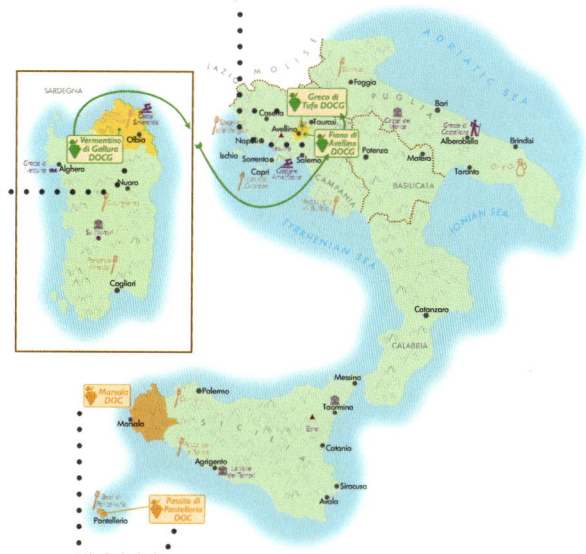

4 **PASSITO DI PANTELLERIA DOC**
Pah-SEE-toh dee Pahn-tay-lay-REE-ah

5 **MARSALA DOC** *Mahr-SAH-lah*

The region: SICILIA/SICILY
See-CHEE-lee-ah

Sicilia is Italy's largest region by size. The Strait of Sicily separates it from the northern coast of Africa, and indeed, the western portion of the island receives quite a bit of influence from the warm Sirocco winds blowing north from that continent. There are several smaller islands off Sicilia's coast that are considered to be part of the larger island and are included in various Sicilian provinces. The volcanic island of Pantelleria is one such example.

Not to miss travel sites:

Palermo, Sicilia's capital, is the island's best-known city. Rich in monuments, art, and architecture, the city is a feast for the eyes. The Cathedral of Palermo houses the remains of Frederick II, Holy Roman Emperor, while the Cappella Palatina chapel is famous for its byzantine architecture and astonishing mosaics. The Teatro Massimo Vittorio Emanuele di Palermo opera house is a treat for music lovers and the historic Mercato di Ballarò food market an important landmark for foodies.

VERMENTINO DI SARDEGNA DOC

Vermentino performs well all over Sardegna and the Vermentino di Sardegna DOC covers the entire island. Vermentino di Sardegna is generally considered a lighter and less-structured version of Vermentino. This style is steadily growing in reputation and renown.

OAK FORESTS

Italy is an important source of cork (sughero). The country's cork production is predominantly concentrated in Sardegna, specifically in Gallura which has large cork-oak forests. The small town of Calangianus (in the middle of Gallura) is considered the Italian capital of cork.

VERMENTINO DI GALLURA DOCG
Sardegna

The Vermentino grape delivers its most prominent expression on the island of Sardegna, especially in the area of Gallura, which is home to the Vermentino di Gallura DOCG.

THE PLACE

The Gallura sub-region occupies the northeastern corner of Sardegna. It has a rugged and rocky landscape characterized by inland woods and olive groves that give way to more typical Mediterranean vegetation along the coast. However, Gallura's strongest geological feature is its granitic mountains and hills. Unsurprisingly, the vineyards here are planted in weathered granitic sands. This soil type happens to be ideal for the Vermentino grape as is the warm, sea-influenced climate.

GRAPE VARIETIES

Vermentino is an aromatic, exuberant grape variety that crafts Sardegna's flagship white wine. But as this grape variety flourishes near the sea, it is also successful in the coastal areas of Liguria and Toscana.

SARDEGNA

Zuppa Gallurese
Costa Smeralda
Vermentino di Gallura DOCG
Olbia
Grotta di Nettuno
Alghero
Nuoro
Culurgiones
Su Nuraxi
Porcetto Arrosto
Cagliari

WINE PROFILE

aromatic, exuberant

What makes this wine special: Gallura produces the most intense, richest and mineral-laden expression of Vermentino. The best wines from this area often show a distinctive briny-salty character not achieved elsewhere.

Acidity: Crisp acidity

Body: Medium- to full-bodied, medium to high in alcohol

Common descriptors: Citrus, ripe apple, stone fruit, tropical fruit, aromatic herbs (sage, rosemary, lavender), Mediterranean scrub, mineral, saline

Food pairings: Appetizers, shellfish, seafood, lobster, fish soup, pasta, white meat

PRODUCTION NOTES

Min 95%, most:100pc

- Vermentino di Gallura requires at least 95% Vermentino, but most producers use 100%
- Vermentino di Gallura Superiore is made with stricter production rules (such as a higher minimum alcohol level)
- Most Vermentino di Gallura is unoaked; however, some producers use oak for their top wines
- Most Vermentino is made in a dry style

NOTABLE PRODUCERS

VERMENTINO DI GALLURA

Capichera: Pioneer producer of top-quality, long-lived Vermentino and the first to make an oaked version

Siddùra: Dedicated to producing wines that work to restore the reputation of Sardinian Vermentino to its esteemed historic status

VERMENTINO DI SARDEGNA

Pusole: Small, family-run winery whose holistic approach to farming and winemaking produces rich, nutty Vermentinos with a saline minerality; bottled under the Ogliastra IGT

Antonella Corda: Corda is a young winemaker who has recently taken over her family estate and is focused on sustainable farming; she crafts fragrant and intense interpretations of Vermentino

SARDEGNA'S SIGNATURE CUISINE

BREAD

Zuppa Gallurese/Suppa Cuata
Zuppa Gallurese is a perfect example of la cucina povera (the food of the poor). The name is slightly misleading as the final product resembles a lasagne more than a soup. This traditional dish combines layers of stale bread, meat broth and pecorino cheese, which are then baked into a sort of deliciously savory bread pudding. Regional variations abound and nutmeg, mint, fennel and onion are all common additions. This ultimate comfort food is traditionally served at Sardinian wedding celebrations.

MEAT

Porcetto Arrosto finds its origins in the island's pastoral traditions. Shepherds would roast a suckling pig over a fire containing aromatic elements such as myrtle, rosemary and juniper. Today, the dish is a staple at weddings and celebrations. In order to ramp up the "delicious factor," the pig can also be stuffed with herbs and meat before cooking. The pig is roasted for about seven hours and is praised for its succulent meat and crispy skin.

PASTA

Culurgioni are a stuffed pasta typical of Sardegna. Similar in appearance to Chinese dumplings, the stuffing is made of potatoes, pecorino cheese and mint. This weighty dish pairs well with the richness of Vermentino.

FIANO DI AVELLINO DOCG
Campania

Fiano di Avellino and Greco di Tufo have both earned recognition as the most distinctive whites of Campania. Although Fiano di Avellino is less well known than its more famous neighbor, Fiano has gained a reputation as one of Italy's noblest and most refined white wines.

in this area Fiano, Greco, Taurasi

THE PLACE

The Fiano di Avellino DOCG is in Irpinia, just to the south of the Greco di Tufo appellation. It covers the area around the city of Avellino, hence its name. (Irpinia is discussed in more detail in the Greco di Tufo section.)

GRAPE VARIETIES

Fiano is a noble, ancient grape grown throughout Campania. Its origins, like those of Greco, have also been traced back to Irpinia. The volcanic soils of the region seem particularly well suited to Fiano and it is on such soils that it produces its most distinctive expressions.

FURTHER EXPLORATION

Falanghina

Falanghina is Campania's most widely planted white grape and is used to produce very distinctive wines. Although Falanghina can craft wines ranging from simple to complex, all are full flavored with stone fruit and tropical fruit balanced by a lively and refreshing acidity. Several appellations make varietal Falanghina in Campania; among the most well known are Falanghina del Sannio DOC, Campi Flegrei DOC and Falerno del Massico DOC.

Mastroberardino

Antonio Mastroberardino was a historic Campanian producer who pioneered the production of quality wine in the region. He is rightly considered the founding father of Campania's modern wine industry. Mastroberardino is also credited with having saved the Fiano grape from extinction, as well as having brought attention to local native grapes such as Greco and Aglianico.

What makes this wine special: Fiano arguably makes some of Italy's longest-lived and most complex white wines that can stand among Italy's most distinctive

Acidity: Pronounced

Body: Medium-bodied, medium in alcohol

Common descriptors: Citrus, floral, pear, apple, herbs, hazelnut, mineral; the wines develop notes of toasted hazelnut, honey, beeswax and petrol with bottle age

Food pairings: Burrata, mozzarella di bufala, charcuterie, seafood, shellfish

PRODUCTION NOTES

- Fiano di Avellino must be made from at least 85% Fiano; however, wines are usually 100% Fiano

- Most wines are unoaked

NOTABLE PRODUCERS

Mastroberardino, Terredora and **Feudi di San Gregorio:** all produce exceptional examples of Fiano di Avellino

Joaquin: Owner Raffaele Pagano crafts expressive, nuanced and textured wines from some of the area's oldest pre-phylloxera Fiano vines

Ciro Picariello: Low-interventionist producer who holds back a portion of bottles each year to be released at a later date. His wines drink well in their youth but can age beautifully for 10-15 years.

Colli di Lapio: Famous for their Romano Clelia Fiano di Avellino, the estate is situated in what is considered the "grand cru" area for the Fiano variety

CAMPANIA'S SIGNATURE CUISINE

PASTA

Spaghetti alle Vongole
Fiano di Avellino is considered the perfect pairing for one of Campania's most classic and well-known dishes, spaghetti alle vongole (spaghetti with clams). This very simple recipe (clams, olive oil, parsley and garlic) is deceptively delicious!

LEMONS

Some of the world's most famous lemon groves hug the coastline of Campania. Several varieties are grown but each produces big, juicy, sweet, delicious fruit. Many of the lemons are used to create the region's famous limoncello (a sweet and intensely lemon-flavored liqueur), while others are incorporated into classic dishes such as insalate di limone (lemon salad) or spaghetti al limone (spaghetti with lemons.)

GRECO DI TUFO DOCG
Campania

Greco di Tufo is one of the most well-known white wines of southern Italy and the best examples are considered among Italy's finest.

Greco vs. Fiano:
most producers in the area will
make both. Greco maybe
more structured, more ageing
Greco, fuller bodied

THE PLACE

Greco di Tufo DOCG lies within the winegrowing district of Irpinia, a historic sub-region of Campania corresponding to the province of Avellino.

Irpinia is qualitatively Campania's most important wine area as it is home to three of the region's flagship wines: Greco di Tufo, Fiano di Avellino and Taurasi. The winegrowing district lies in the hills and mountains of central-east Campania.

Many of the vineyards are planted on slopes, often at significant elevations. Soils range from a mixture of limestone, sand and clay (of marine origin) to fertile soils of ash, pumice and lapilli (as a result of recent volcanic activity).

Particularly unique are the sites that possess significant sulfur deposits which craft full-bodied, mineral-driven wines. As opposed to other southern Italian winegrowing areas, vineyards here benefit from a somewhat cooler climate and a longer growing season. These unique conditions make the region ideally suited for the production of high-quality, complex white wines.

GRAPE VARIETIES

Greco is one of Campania's most ancient grapes and among its finest. Although it is grown across the region, its true home is the area surrounding the Irpinian village of Tufo.

GRECO

A grape from Greece?
There are several theories surrounding the origin and name of the Greco grape. Many presumed that the grape was brought to southern Italy nearly three thousand years ago by the ancient Greeks. However, no genetic relationship has been found between Greco and modern-day Greek varieties.

Alternatively, the name could have originated in reference to the highly reputed, Greek-styled sweet wines that were popular in ancient times. The grapes that were used for the production of wines made in the "Greek" style, ended up being named after the style of wine they produced.

What makes this wine special: Greco di Tufo is considered the consummate expression of the Greco variety. The best wines show firm structure, minerality and a textured mouthfeel, all sustained by elevated acidity.

Acidity: Pronounced

Body: Full-bodied, medium alcohol

Common descriptors: Citrus, pear, quince, stone fruit (peach), tropical fruit, hazelnut, mineral; with bottle age the wines develop complex notes of honey and toasted nuts

Food pairings: Mozzarella di bufala, charcuterie, seafood (grilled tuna), pasta with cream sauce, risotto, grilled vegetables

PRODUCTION NOTES

min: 85%
most 100%

- Greco di Tufo must include at least 85% Greco; most usually contain 100% Greco

- Most wines are unoaked

NOTABLE PRODUCERS

Mastroberardino: This estate was founded by Antonio Mastroberardino who worked tirelessly after his return from WWII to restore his family's vineyards and promote the wines of Campania

Terredora Di Paolo: Walter Mastroberardino (Antonio's brother) is also a key figure in the Campanian viticultural renaissance; Terredora, his family-owned company, has evolved into one of southern Italy's largest wineries

Feudi di San Gregorio: Founded in 1986, this winery is credited with garnering world-wide attention for Irpinia with cutting-edge labels, forward-thinking architectural design, avant-garde marketing campaigns, and award-winning wines

Benito Ferrara: Widely considered by many critics to be the top producer of Greco di Tufo, their iconic Greco di Tufo Vigna Cicogna is among Italy's best white wines

Pietracupa: Pietracupa is a young estate that has quickly gained a reputation for producing some of the most pure and mineral Greco di Tufos on the market

Quintodecimo: Luigi Moio is one of Campania's most gifted winemakers. He crafts exceptionally refined Greco from high-elevation vineyards.

CAMPANIA'S SIGNATURE CUISINE

CHEESE

Mozzarella di Bufala (buffalo mozzarella) is one of Campania's specialties and pairs nicely with Greco di Tufo. This creamy, soft cheese is made from the fresh milk of the local Italian Mediterranean buffalo (a breed of water buffalo). Buffalo mozzarella has a higher fat and protein content and is more intensely flavored than cow's milk mozzarella.

Insalata Caprese (Caprese salad) is a much-beloved vehicle for the region's mozzarella. It is often served at the start of a meal and pairs the fresh cheese with tomatoes, sweet basil, olive oil and a sprinkle of salt. True Campanian buffalo mozzarella is sold exclusively under the designation of the Mozzarella di Bufala Campana DOC.

CHESTNUTS

Castagna di Montella were among the first Italian fruits to be recognized with a designation of origin. The famous chestnuts are grown near the town of Montella in Irpinia's Alta Valle del Calore. In the past, the chestnut tree was often referred to as the "bread tree," as the nut flour it provided was considered essential for survival for many of the region's mountain inhabitants. Today, there are festivals devoted to this delicious nut and castagne can be enjoyed fresh, roasted, cooked, dried and ground into the famous life-sustaining flour.

PASSITO DI PANTELLERIA DOC
Sicilia

For more than 2,000 years, Zibibbo grapes have been transformed into lusciously sweet elixirs on the volcanic island of Pantelleria. Today, the Passito di Pantelleria DOC is responsible for producing some of the world's finest sweet wines. Crafted entirely from Zibibbo, the wines are said to capture the very essence of this special island.

Giardino Pantesco
With virtually no rainfall or groundwater and almost constant gale-force winds, agriculture on Pantelleria proves challenging. Islanders have devised ingenious ways to shelter plants from the extreme elements. One of the most amazing of these systems is the giardino pantesco. Often constructed around a single citrus tree, these circular structures were designed at a height that would block winds, while still allowing the perfect amount of sunshine to reach the interior. Inward sloping walls, built of lava stone, captured every drop of dew and fog available.

Dammusi
The island of Pantelleria is dotted with traditional dwellings known as dammusi. Constructed with thick lava-rock walls, the structure helped to keep occupants cool in the summer and warm in the winter. The domed roof served to channel the island's sparse rainfall into cisterns below and also reflects the Arabic influence in its esthetic. Today, many dammusi have been renovated and are available to rent as accommodation for the island's visitors.

THE PLACE

The name **Pantelleria** is derived from the Arabic phrase "Bent El-Rhia," or "Daughter of the Wind." Situated off the coast of Sicily's westernmost tip, this "Black Pearl of the Mediterranean" is the largest of Italy's volcanic satellite islands. On a clear day, the coast of Tunisia can be seen from its shores—which comes as no surprise as the island is actually closer to Africa than to Italy.

The ever-present sirocco and mistral winds that collide over Pantelleria have shaped both the landscape and the inhabitants of the island. According to Greek mythology, Pantelleria has been bewitching people and gods alike for thousands of years; the goddess Tanit is said to have seduced Apollo with the island's enchanting Moscato wine.

GRAPE VARIETIES

Pantelleria is crafted from Zibibbo (*Zee-BEEB-boh*), also known as Moscato di Alessandria. The variety was brought to the island by Arabs and its synonym hints at its proposed Egyptian origins. Originally, it was planted to produce table grapes and raisins. In fact, the name Zibibbo comes from the Arabic word "Zabīb" meaning "dried grape."

WINE PROFILE

basically little vine shrubs (handwritten)

What makes this wine special: These wines are the product of "heroic" viticulture. Rainfall is scarce, groundwater almost non-existent and constant winds assault the vines. This harsh environment necessitates that the vines are grown low to the ground. In fact, they are actually planted in shallow depressions that help protect the plants and conserve moisture. Grapes must be picked by hand and the winemaking process is labor intensive. The resulting wines are rich, full-bodied and lusciously sweet. Well-made examples feature balancing acidity that hints at the saline sea-spray that buffets the island's shores. The wines are capable of aging and develop lovely nutty, chocolatey notes with time.

Acidity: Balancing

Body: Full-bodied

Common descriptors: Apricot, date, marmalade, sage, candied fig, orange blossom, candied peach and lemon

Food pairings: Blue cheese, aged cheeses, chocolate, and pastries such as baci panteschi

PRODUCTION NOTES

they will have high levels of acidity (handwritten)

- An early harvest of healthy grapes occurs toward the end of August. The grapes are laid out on racks and left to wither naturally in the sun and wind for up to a month.
- A second harvest is conducted in September and the must from this harvest is allowed to begin fermenting right away *these will have ↓ acid ↑ sugar* (handwritten)
- During fermentation, the dried grapes (from the first harvest) are added in several lots to the fermenting must of the second harvest
- The dried grapes macerate in the fermenting must, releasing flavor, color, sweetness and acidity
- The fermentation continues slowly for over a month, or until the wine achieves a balance of sweetness and acidity
- Aging regimes vary from producer to producer, but the wines must age a minimum of six months before release

NOTABLE PRODUCERS

De Bartoli: Marco de Bartoli is a historic Marsala producer who recognized the significance of the wines of Pantelleria decades ago. He bottled his first Passito di Pantelleria, Bukkuram, in 1984. Bukkurum, in Arabic, means "father of the vineyard" and honors the part of the island that was favored for the production of grapes in ancient times.

Donnafugata: The producer of the iconic Ben Ryé Passito di Pantelleria, often cited as one of the world's best sweet wines

Pellegrino: A large producer that has been instrumental in introducing the world to Passito di Pantelleria

Salvatore Murana: One of the top producers of Passito di Pantelleria

Fabrizio Basile: Innovative producer that is known for wines aged in chestnut and acacia

Ferrandes: Organic estate that works with old vines planted in 1930 and 1965

PANTELLERIA'S SIGNATURE CUISINE

The cuisine of Pantelleria is referred to as "pantesco." Pantesco fare is unique in that, despite the island's Mediterranean location, seafood plays a very limited role. The residents have traditionally focused on farming, not fishing. One of the island's most popular dishes is couscous, or in the original Arabic, kuskus. Traditionally, the kuskus would be served with a selection of the island's vegetable produce.

CAPERS

Pantelleria is home to some of the world's best capers (unopened flower buds of the caper bush). Caper bushes can be found growing wild throughout the island but are also cultivated by farmers. If the buds are not picked, a beautiful flower will bloom and develop into a large caperberry.

DESSERT

Baci di Pantelleria are a decadent island treat. Special molds are used to produce a flower-shaped pastry fritter. Two fritters are sandwiched together with a generous dollop of ricotta in the middle. Dusted with icing sugar and drops of chocolate, this traditional pastry is the perfect pairing to Passito di Pantelleria.

MARSALA DOC
Sicilia

Marsala was once widely considered one of the world's greatest fortified wines. Hugely popular in the 18th and 19th centuries, Marsala's commercial success made wine a key facet of the Sicilian economy. However, Marsala became the victim of its own popularity and quality began to falter by the second half of the 20th century. In an effort to bring the appellation back to its former glory, regulations were introduced in the 1980s that focused on increasing wine quality. Today, the top Marsala producers are crafting some of the world's more distinctive and delicious wines.

FURTHER EXPLORATION

John Woodhouse was an English trader who is credited with accidentally discovering Marsala. Legend has it that his ship was forced to take shelter from a storm in the port of Marsala. Stranded in the city, he began to seek out the area's best refreshments. The wine that he was served reminded him of wines such as Port and Madeira. Confident that the wines would be greatly appreciated in his home country, he purchased 50 barrels of the wine. As a safety measure, he added brandy to the barrels to increase their stability during the long ocean voyage. Woodhouse renamed the tipple Marsala and its reception in England was so successful that Woodhouse moved to Marsala to devote his life to the business. He and his brothers set a wine trade in motion that eventually led to Marsala being exported around the world.

THE PLACE

The Marsala DOC occupies the western corner of Sicily. This is the island's most productive winegrowing area and is one of the most densely planted winegrowing areas in all of Italy. The DOC covers a large area that is diverse in topography and soil types, but homogenous in climate—this is one of the warmest and driest parts of mainland Sicily. Most of the appellation's vineyards lie on coastal plains and low-elevation hills further inland.

GRAPE VARIETIES

Though other grapes are permitted, Grillo *(GREEL-lo)* and Catarratto *(Ka-tar-RAT-toe)* are the most important varieties for Marsala production.

Grillo: This native Sicilian grape is considered to make the highest-quality Marsala. It contributes acidity, texture and aromas to the final wine. It is also capable of ripening to high sugar levels.

Catarratto: Although traditionally favored because of its higher yields, Catarratto does not ripen to sugar levels as high as Grillo, therefore Catarratto-based wines require more alcohol during fortification. Catarratto also has a tendency to oxidize quickly, adding oxidative flavors and a darker color to the wines.

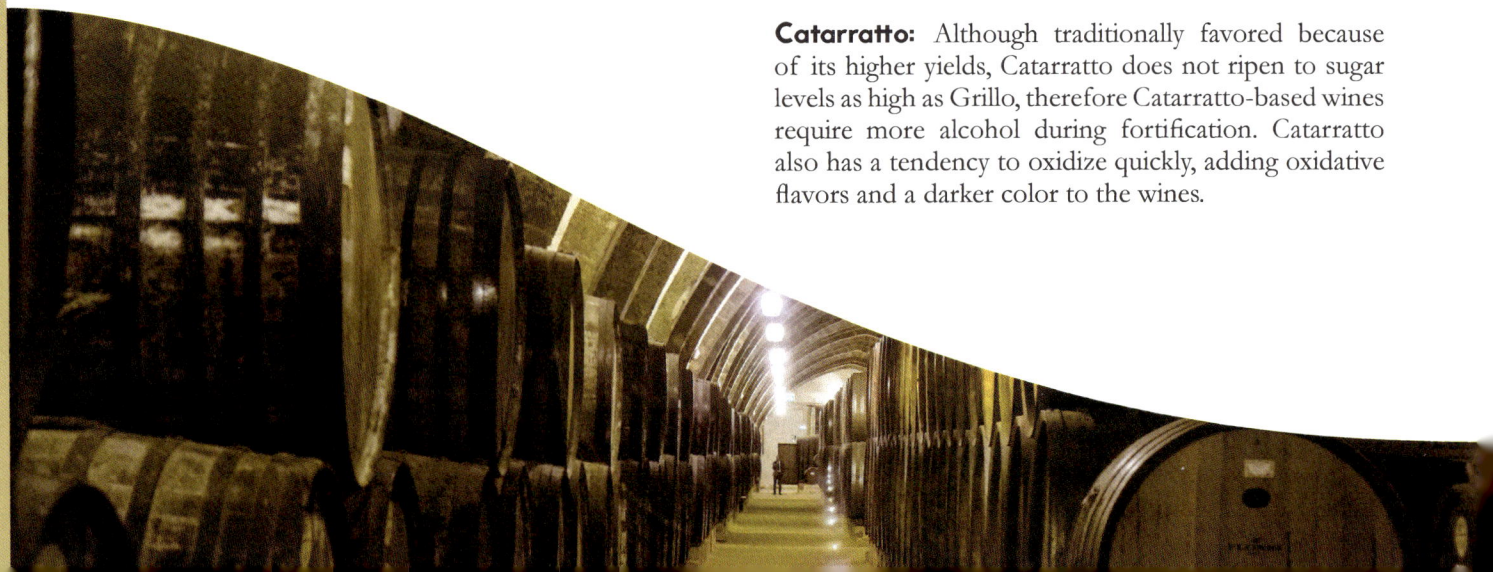

Many variations: colour : amber, ruby
sweetness : dry to sweet
ageing

WINE PROFILE

What makes this wine special: Marsala is a fortified wine. This is to say that neutral grape spirit is added to the wine to boost its alcohol level. Some of the best examples of Marsala are aged in perpetuum, i.e. casks of aged wine are never emptied completely. Only a portion of the wine is removed from any given cask (for bottling) and then this void is replaced with younger wine, which the older wine assimilates.

Acidity: The best examples have fresh acidity *to balance the sweetness*

Body: Full-bodied

Common descriptors: Sweetness levels vary greatly but the best wines will share similar flavor characteristics—notes of walnut, date, black tea, tobacco, treacle, spice, almond, sea salt, caramel, honey, coffee and resin

Food pairings: Drier versions pair beautifully with hard or spicy cheeses, while sweeter versions are a natural match to desserts. The very best examples are termed vino da meditazione (meditation wine) and thought to be best enjoyed without food.

NOTABLE PRODUCERS

Marco De Bartoli: Credited with preserving the historic honor of the Marsala DOC; the estate remains a standard bearer for top-quality Marsala

Cantine Florio: Florio was the first Italian producer of Marsala wine; the estate has changed hands several times, yet it still remains the largest producer of Marsala

Pellegrino: Founded in 1880 by Sicilian Paolo Pellegrino, this producer has been pivotal in expanding Marsala exports to foreign markets

Curatolo Arini: This is the oldest family-owned Marsala company and is now under the leadership of the fourth generation. Their distinctive labels by Sicilian architect Ernesto Basile have remained unchanged since their initial design more than 100 years ago.

PRODUCTION NOTES

Styles of Marsala are determined by what is added to the wine and how the wine is aged. The wines can be broadly divided into two categories: Marsala Vergine and Marsala "conciati".

- **Marsala Vergine** is a dry wine that is made from white grapes only and must be aged for at least five years in partially filled barrels. Aside from the spirit added at the time of fortification, no other additions can be made. If Marsala Vergine ages for at least 10 years in barrel, it is referred to as riserva or stravecchio. *pure*

- **Marsala "conciati"** allows for additions such as cooked grape must, fortified grape must or grape spirit. These embellishments are used to bolster the color, flavor and/or mouthfeel of the wine. *with additions*

Marsala Superiore and Marsala Fine are both examples of Marsala "conciati" and are made in many styles with a wide range of sweetness levels and aging regimens.

- **Marsala Superiore** accounts for approximately 20% of Marsala production. Some of the finest examples are found within this category. The wines must mature in wooden barrels for at least two years and can be released as riserva after four years of wood maturation.

- **Marsala Fine** accounts for almost 80% of the total Marsala production. The majority is intended to be used as a cooking ingredient and is sold to the food industry. Marsala Fine must be matured for a minimum of one year, with eight months in wood. *for cooking*

SICILIA'S SIGNATURE CUISINE

DESSERT

Cannoli Siciliani is a true melding of Arab and Christian cultures. It is said that the delicacy was developed by women of the Emir's harem during the period of Arab domination. When Arab rule ended, many of these women remained in Sicilia and sought refuge in monasteries, eventually converting to Christianity. The women are said to have handed down the recipe for cannoli to the Christian nuns who began to bake the treat for carnival season. Today cannoli are a staple of Sicilian cuisine. This delicious pastry consists of tube-shaped shells of fried pastry dough, filled with a sweet, creamy filling usually containing ricotta, a touch of sweet Marsala and sprinkled with chocolate or pistachios and candied fruit.

NOTES

SOUTHERN ITALY
Red Wines

SOUTHERN ITALY RED WINES

0   30   60   90 km
0   25   50 miles

SARDEGNA

Costa
Smeralda

Casu
Marzu

Olbia

*Grotta di
Nettuno* ● Alghero   *Su Filindeu*

Nuoro

*Pane
Carasau*

Su Nuraxi

**Cannonau di
Sardegna DOC**
covers the entire island

Bottarga

Cagliari

LAZIO   M O L I S E

*Pane di
Altamura*

Foggia

P U G L I A

Bari

*Reggia di
Caserta*

*Pizza
Napoletana*

Benevento

Caserta   Avellino ● Taurasi

*Castel del
Monte*

*Grotte di
Castellana*

Alberobello

Brindisi

Napoli

Vesuvio
● Pompei

**Taurasi
DOCG**

Ischia   Sorrento

Salerno

Potenza

Matera

Capri   *Costiera
Amalfitana*

C A M P A N I A

BASILICATA

Taranto

**Primitivo di
Manduria
DOC**

*Orecchiette
pasta*

*Spaghetti alla
Puttanesca*

T Y R R H E N I A N   S E A

I O N I A N   S E A

A D R I A T I C   S E A

Catanzaro

CALABRIA

Messina

Granita

*Pasta alla
Norma* ● Palermo

Taormina

S I C I L I A

Marsala

*Pistacchio
di Bronte*

Etna

**Etna
DOC**

Catania

Agrigento

*La Valle
dei Templi*

Siracusa

Avola

**Sicilia Nero
d'Avola DOC**
covers the entire island

*Pantelleria
Capers*

Pantelleria

Map by
Quentin Sadler
WINE
SCHOLAR
GUILD

1 CANNONAU DI SARDEGNA DOC
Kahn-NOH-now dee Sahr-DEH-nyah

The region: SARDEGNA/SARDINIA
Sahr-DEH-nyah

The island of Sardegna occupies a key strategic position in the Western Mediterranean and has endured many conquests. Despite past periods of occupation, the Sardi people have managed to maintain a strong and independent spirit.

The island's inhabitants have historically been farmers and hunters, and the vine has been part of Sardinian culture since ancient times.

Sardegna has a rich ampelographic heritage of its own, but both Spanish and mainland Italian varieties are well represented. The island boasts one of the highest concentrations of centenarians in the world, a phenomenon, at least partially, attributed to the island's wine!

Not to miss travel site:
Cagliari, a coastal city in the south, is Sardegna's capital and the island's most important port. History buffs are drawn to the Museo Archeologico Nazionale di Cagliari which houses an incredible collection of artifacts from the Nuragic age to the Byzantine era.

2 SICILIA NERO D'AVOLA DOC
NEH-roh DAH-voh-lah

3 ETNA DOC
EHT-nah

The region: SICILIA/SICILY
See-CHEE-lee-ah

The Strait of Messina separates Sicilia from Calabria—1.9 mi/3.1 km at its narrowest - but Sicilia maintains a distinct identity from mainland Italy. Modern Sicilian culture reflects its past invaders and settlers, but remains uniquely Sicilian.

Sicilia's terrain varies widely, but hills are common throughout the island, and indeed, most vineyards are planted on hillsides. The Sicilian Apennines are an extension of the Southern Apennines and include Monte Etna, Europe's largest active volcano.

SOUTHERN ITALY RED WINES:
THE ROAD MAP

Not to miss travel sites:
The town of Taormina sits atop a cliff, overlooking the Ionian Sea on Sicilia's east coast. Aside from the breathtaking views, the city's historic and archaeological heritage has made this one of Italy's most popular destinations. Monte Etna provides a dramatic backdrop for this charming town and is an attraction in its own right. Designated as a UNESCO World Heritage Site, this active volcano is Italy's highest peak south of the Alps. The possibility of eruption has not deterred tourists who flock to visit its remarkable crater.

Sicilia is renowned for a rich treasury of historic architecture and archaeological sites seemingly found in every corner of the island. La Valle dei Templi di Agrigento (Valley of the Temples) is one the world's best remaining examples of ancient Greek temples. Multiple cities in the island's southeastern corner are so famed for their Baroque buildings that the area is referred to as "Sicilia Barocca."

4 TAURASI DOCG *Tah-RAH-zee*

The region: CAMPANIA
Kahm-PAH-nee-ah

Small, twisty roads leading to medieval hillside villages are a hallmark of Campania, but these villages are anything but remote. Campania has Italy's highest population density and Naples, its capital, has Italy's third-largest number of inhabitants behind Rome and Milan.

Not to miss travel site:
Napoli is one of Europe's oldest cities and it has a wealth of historic attractions to visit. Centro Storico, the earliest section of the city, has preserved the ancient Greek grid layout which continues to be mirrored in many modern cities. The Piazza del Plebiscito is the city's most important square and is ringed with important landmarks such as the San Francesco di Paola church, the staggering Palazzo Reale (Royal Palace), and the Teatro di San Carlo, the world's oldest operating opera house. Even the ground below the city streets offers an enchanting view of history. Centuries of excavation and mining have resulted in a complex network of subterranean caves and tunnels that is almost a city of its own and includes a series of much-visited catacombs.

DETOUR TO PRIMITIVO DI MANDURIA DOC
Pree-mee-TEE-voh dee Mahn-DUR-ee-ah

The region: PUGLIA/APULIA
POO-lyah

Puglia represents the "spur" and "heel" of Italy's boot-shaped landmass. This is Italy's least mountainous region. Puglia's topography is dominated by large plateaus, fertile plains and low hills.

Volumetrically, Puglia competes with Veneto and Emilia-Romagna as Italy's largest producer of wine. In the past, it was common for the rich, dark, full-bodied wines of Puglia to be brokered on the bulk market and blended with less-ripe wines from Europe and the northern regions of Italy. Yet, Puglia, blessed with a host of high-quality native grapes, is quite capable of crafting distinctive wines of quality and character. It is one of Italy's major producers of rosato, and without question, some of the country's best examples are produced here.

CANNONAU DI SARDEGNA DOC
Sardegna

SARDEGNA

Cannonau is the island's flagship variety and, though the grape is planted elsewhere in Italy, it shines brightest here. Sardinian wines crafted from Cannonau have received world-wide attention due to their high levels of antioxidant compounds, thought to contribute to the longevity of the island's peoples.

Sardinia's most planted variety

Cannonau di Sardegna DOC covers the entire island

THE PLACE

The mountains of Sardegna no longer reach the heights they once did. Weathered and eroded over millennia, most of the island's peaks are now rugged hills intermingled with rocky uplands and plateaus. The Gennargentu Mountain range on the east-central portion of the island retains the highest peaks—these give way to a narrow band of rocky, craggy hills towards the east coast and lower-elevation hills and a smattering of plains to the west.

Cannonau is planted across the island and most is bottled under the regional Cannonau di Sardegna DOC. However, the greatest concentration of vines can be found in the Nuoro province in the east-central part of the island. Here, warm, dry conditions and weathered granitic soils help the grapes to ripen to high sugar levels, while still retaining balancing acidity.

GRAPE VARIETIES

Cannonau has been thriving in Sardegna for more than 400 years and, as the island's most widely planted variety, is an integral part of the island's viticultural landscape. DNA testing has proven that Cannonau is identical to Garnacha, a grape historically thought to be from Spain, but scholars continue to debate the variety's true country of origin.

NURAGHI

The remnants of more than 7,000 megalithic stone structures known as "nuraghe" (nuraghi is the plural) are scattered throughout Sardegna's interior. These Bronze Age monuments were constructed with dry stones in an unparalleled architectural style. While the engineering is fairly well understood, their purpose or function is the subject of much debate. Some experts propose that they were used as defensive structures, others argue that they were religious in purpose, or even built as status symbols. Regardless, the Nuraghi have become an intrinsic part of Sardinian culture. The archeological site of Su Nuraxi di Barumini, in south-central Sardegna, is the most complete example of this prehistoric architecture and has been classified as a UNESCO World Heritage Site.

CONSORZIO DI TUTELA VINI DI SARDEGNA

WINE PROFILE

What makes this wine special: Cannonau has the ability to produce wines that range from light-bodied, fruity, and easy-drinking to structured, complex, and age-worthy. The wines of Cannonau di Sardegna DOC are often marked with nuances of Mediterranean scrub and a salty-briny note that differentiates them from Spain's Garnacha. Additionally, wines coming from the granite soils around Nuoro can display a stony, mineral quality that is quite distinctive.

Acidity: Low to moderate

Tannin: Low to high, depending on vineyard site and winemaking

Body: Light- to full-bodied, with generous alcohol

Common descriptors: Red cherry, raspberry, plum, meadow flowers, dried herbs, white pepper. With age, the wines can develop notes of red licorice and spice.

Food pairings: Cannonau di Sardegna pairs well with roasted meats and vegetables, as well as Fregula, a Sardinian pasta similar to couscous. Cheeses such as Fiore Sardo and Pecorino Sardo also complement the wine nicely.

NOTABLE PRODUCERS

Dettori: The vineyards of Dettori are farmed organically and the estate's wines are bottled unfined, unfiltered and often without sulfur additions; offers some of the island's best expressions of Cannonau

Argiolas: One of the first Sardinian producers to shift focus from quantity to quality; they remain a benchmark producer

Perda Rubia: Historic winery that has focused on Cannonau from their inception; the grapes for their most iconic wine, Naniha, are sourced from ancient, pre-phylloxera vines

Pala: This award-winning family-owned winery farms organically in order to produce terroir-driven bottlings

PRODUCTION NOTES

Min 85%
most: 100%

- Cannonau di Sardegna DOC requires a minimum of 85% Cannonau; most wines are 100%

- Riserva versions must be aged for two years, with at least six months in wood

- Wines can be labeled classico if made in one of the more restricted zones of production (among several other criteria)

SARDEGNA'S SIGNATURE CUISINE

BREAD

Pane Carasau, a paper-thin crispy bread, is a specialty of Sardegna. Locally, it is referred to as Carta da Musica (sheet music). Many believe the bread earned the nickname because its texture is similar to parchment paper (on which music was traditionally printed); others say it is because the bread was to be made thin enough so that sheet music could be read through it. This wafer-thin bread is baked in large circles, sold in stacks and cut into triangles before serving. It stays fresh for weeks and is best enjoyed with a slice of fresh pecorino and olive oil.

FISH

Bottarga, or salted roe, has been enjoyed for more than 5,000 years and is one of the most ancient processed foods. Sardegna is famous for the quality of its mullet fish bottarga and the most prized is sourced from fish that live in the Pond of Cabras. After the fish are harvested and the egg sacs carefully removed, the pouches are brined, then packed in dry salt and pressed for 40 days. During this time, liquid is slowly extracted from the roe. The brined roe is then hung to dry for at least another month. Known as Oro di Cabras (Gold of Cabras), this golden delight is said to be the smoothest-textured bottarga in Italy. The unique flavors of this umami-rich, salty delicacy are best highlighted by serving it with bread that has been soaked in olive oil or by shaving it over simple pasta dishes

SICILIA NERO D'AVOLA DOC
Sicilia

Nero d'Avola is Sicilia's most prolific red grape variety and a cornerstone of most of its red blends. Although a great deal of Nero d'Avola is bottled under the regional Terre Siciliane IGT, an increasing amount is being bottled under the flexible, regional Sicilia DOC.

No needs heat - retaining dark soils doesn't mind drought-like conditions

THE PLACE

Sicilia Nero d'Avola DOC wine is produced throughout Sicilia, but the island's southeastern corner produces some of the most captivating examples. Indeed, the variety is thought to have originated near the coastal town of Avola.

The Eloro DOC, south of Avola, is well regarded for crafting outstanding wines from this grape—especially around the village of Pachino. Nero d'Avola loves heat and the climate of southeastern Sicilia does not disappoint. The variety thrives in the area's heat-retaining dark soils and basks in the warm and arid growing conditions.

GRAPE VARIETIES

Nero d'Avola was historically known as Calabrese and is still cited as such in Italy's National Registry of Grapes. The grape has a reputation for reflecting the terroir of where it is planted and, in the right hands, has the capability of producing complex and compelling wines.

The variety requires warm growing conditions to ripen fully and is often trained close to the ground to absorb the heat reflected back from the sun-kissed earth. Despite its ability to ripen to high sugar levels, Nero d'Avola is able to retain its refreshing levels of acidity even in the warmest weather—a real blessing in Sicilia's hot, Mediterranean climate.

Duomo di San Giorgio, Ragusa

CERASUOLO DI VITTORIA

Cerasuolo di Vittoria DOCG
In 2005, Cerasuolo di Vittoria became Sicilia's first DOCG thanks to the efforts of local producer, COS. The area lies between the sea and nearby Monti Iblei. The appellation is well known for the pockets of reddish, iron-rich topsoil known as "terra rossa" and this unique soil is considered ideal for the Frappato grape. Frappato is historic to the area and makes light-bodied wines that are perfumed, lively, fresh and juicy, with gentle tannins.

Cerasuolo di Vittoria DOCG must contain 50-70% Nero d'Avola and 30-50% Frappato. The wines range from succulent, red fruit-dominant bottlings to those with overt herbal influence and significant tannic grip, depending on the varying percentages of Nero d'Avola and Frappato.

WINE PROFILE

What makes this wine special: Nero d'Avola can be made in a variety of styles. The wines can range from red-fruited, medium-bodied examples to dense, dark and chewy renditions with spice and black fruit. Terroir and winemaking choices greatly affect the resulting wine style.

Acidity: High

Tannin: Dense, smooth and soft

Body: Medium- to full-bodied, typically high in alcohol

Common descriptors: cherry, plum, blackberry, aromatic herbs and underbrush

Food pairings: For lighter examples, pair with rich fish soups, eggplant caponata and grilled vegetables. For denser, more structured wines, roasted lamb, oxtail soup, beef or lentil stew and wild game are recommended.

PRODUCTION NOTES

Sicilia Nero d'Avola DOC

- Wines must be a minimum of 85% Nero d'Avola
- Riserva versions must be aged for two years

WINES of SICILIA DOC
EXPLORE A MOSAIC OF FLAVORS

NOTABLE PRODUCERS

Tasca d'Almerita: One of the most prestigious producers of the Contea di Sclafani DOC. Here, Nero d'Avola is blended with Perricone in the DOC's only permitted red wine, Contea di Sclafani Rosso.

Planeta: One of Sicily's largest vineyard owners; their "Santa Cecilia" bottling shows the potential of Nero d'Avola from the Noto DOC

Gulfi: Certified organic estate specializing in Nero d'Avola. The grapes for their "NeroSanlorè" Nero d'Avola come the historic San Lorenzo vineyard. This vineyard boasts 40-year-old vines and is situated in the revered Pachino cru area.

Donnafugata: One of Sicily's most prestigious wine families; known for their juicy, fresh style of Nero d'Avola

COS: Iconic and revolutionary producer from the Vittoria area of southeastern Sicily; crafts great single-varietal Nero d'Avola (bottled as IGT) and delicious Cerasuolo di Vittoria

SICILIA'S SIGNATURE CUISINE

ARANCINI

It is thought that arancini were invented during the Arab occupation of Sicilia in the 10th century. The name, which is translated as "little orange," derives from their shape and color which is said to resemble an orange. Arancini produced in eastern Sicily have a more conical shape, said to reference the shape of Monte Etna. Regardless of the shape, the balls of saffron-flavored rice are stuffed with a variety of fillings, coated with breadcrumbs and deep-fried. The most popular filling is ragù but many other fillings such as caciocavallo cheese, peas and prosciutto can also be found. Sampling each region's arancini is a delicious way to learn more about local culture!

PASTA

Pasta alla Norma is one of Italy's most adored pasta dishes. Legend has it that this dish, which originated in Catania, is said to have been named in homage to Vincenzo Bellini's opera, Norma. Upon tasting the dish, Italian writer, Nino Martoglio, was said to exclaim "It's a Norma!", equating the dish to the near perfection of Bellini's opera. The original recipe is thought to have been made with maccheroni, tomatoes, fried eggplant, ricotta salata and basil, but each household has their own variation of this delicious pasta.

NERELLO CAPPUCCIO

Nerello Cappuccio was rarely bottled as a varietal wine in the past due to its tendency to produce less complex wines than its more popular cousin, Nerello Mascalese. Instead, it was traditionally used as a blending wine, adding color and softening the tannic grip of Nerello Mascalese. Historically, the varieties were co-planted, and this is still the case in some of Etna's oldest vineyards. Today, winemakers are beginning to recognize Nerello Cappuccio's potential for producing fruity, refreshing, easy-drinking wines and varietal bottlings are becoming more common.

CONTRADA

The contrade (plural) of Etna have been used for generations to describe and differentiate between various territorial sub-divisions. While many of the contrade are defined by the lava flows that have geologically shaped the landscape, others have developed based on the traditional use of the land.

Not unique to Etna, contrade exist throughout Sicilia and it is not unusual for vineyards to be identified by contrada (singular), rather than by vineyard names. The Etna DOC is the only Sicilian wine region allowed to use the names of the contrade on wine labels.

ETNA DOC
Sicilia

Now one of the most fashionable vineyard areas in Europe, the magnificent vineyards of Etna were once virtually unknown outside of Sicilia. Near the turn of the 21st century, a handful of local producers and winemakers from mainland Italy became focused on revitalizing and nurturing the region's potential. Since that time, Etna has seen a flurry of investment and new vineyard plantings, resulting in critical acclaim for its wines. The sheer variety of soil types, vineyard sites and meso-climates makes the region a treasure trove for terroir hunters.

THE PLACE

Monte Etna, or as the locals affectionately call it "Mongibello" (Beautiful Mountain), dominates the landscape of northeastern Sicilia. Recognized as a UNESCO World Heritage Site in 2013, this ancient and active volcano looms large over the province of Catania and provides fertile ground for some of Europe's highest-elevation vineyards. There are definitely alpine influences in spots, with snow a common sight during winter months.

The Etna DOC wraps around the northern, eastern and southern slopes of the volcano, almost like a backward "C." The frequent eruptions of Monte Etna have repeatedly altered the landscape; therefore, vineyards are often confined to small, terraced parcels supported by walls built from volcanic rock. The land is split administratively between 20 villages that are further divided into historical sub-divisions known as contrade.

Some of Europe's highest vineyards

GRAPE VARIETIES

Found almost exclusively on the slopes of Monte Etna, Nerello Mascalese (*Neh-REHL-loh MAHS-kah-LEH-zeh*) is considered one of Italy's top red grape varieties. The name references the color "nero" (black) and the town of Mascali on Etna's eastern slopes. Nerello Mascalese is often compared to both Nebbiolo and Pinot Nero for its ability to reflect the nuances of the site on which it is grown. This late-ripening variety needs favorable autumn conditions to ensure that its generous tannins fully mature.

CONSORZIO TUTELA VINI
ETNA DOC

PROFILE

What makes this wine special: Capable of great complexity and elegance, the wines can be magnificent translators of terroir. Aroma and flavor descriptors often lead to comparisons with Pinot Nero, while the structural components are reminiscent of Nebbiolo.

Acidity: High

Tannin: High

Body: Medium- to full-bodied, with high alcohol

Common descriptors: Pale to medium intensity of color with flavors and aromas of sour cherry, aromatic herbs, tobacco, smoke and mineral

Food pairings: Braised meats, eggplant dishes, pork, poultry, oily fish, dishes with tomatoes and peppers

PRODUCTION NOTES

Min 80pc most 100pc

- Etna Rosso must be made from a minimum of 80% Nerello Mascalese. Up to 20% Nerello Cappuccio can be added; however, many wines are pure Nerello Mascalese.
- Most wines are matured in wood. Small oak barrels are common, but large barrels (botti) are increasingly seen. Historically, the local chestnut trees were used to craft these barrels and some producers continue this tradition.
- Etna Rosso Riserva requires a minimum of four years aging with at least one year in oak

NOTABLE PRODUCERS

I Vigneri: A winegrower and producer co-operative founded by one of Etna's most passionate and influential oenologists, Salvo Foti; he is considered by many to be the Godfather of Etna wine

Benanti: A trail-blazing producer whose commitment to Etna and its native grapes set the stage for the region's current fame and fortune

Passopisciaro: Benchmark producer that was pivotal in the re-emergence of Etna as one of Italy's top terroirs

Tenuta delle Terre Nere: Pioneering producer that was one of the first to highlight individual terroirs by bottling single-parcel Nerello Mascalese

Frank Cornelissen: Controversial producer who takes an extreme hands-off approach to farming and winemaking and crafts wines of unique character

SICILIA'S SIGNATURE CUISINE

PISTACHIOS

Pistacchio Verde di Bronte DOP
The famous pistachio nuts of Bronte are widely acknowledged as the world's best pistacchi and are known as the "green gold" of Sicily. The small trees dot the foothills of Etna's western slopes and their unique flavor is largely attributed to the volcanic soils in which they grow. Demand always exceeds supply, and these highly prized nuts often sell for twice the price of pistachios from other regions. Authentic Bronte pistachios have brilliant green flesh, violet-colored skin and a pointed shape. They are very sweet tasting and rarely salted in the shell. As with wine, the genuine article should have the official DOP stamp to guarantee authenticity.

DESSERT

Granita
Every June, the coastal city of Acireale hosts the Nivarata Acireale Granita Festival. The term "nivarata" refers to the name of the men who, in times past, collected and stored Etna's snowy bounty. Snow would be packed in ancient lava tubes around the mountain and then transported down as blocks of ice come summer. The ice would be shaved and turned into tasty treats to help beat the summer heat. The Arab occupation added citrus and sugar to the list of granita flavorings, but the locals got even more creative. Today, granita is the quintessential Sicilian breakfast, especially brioscia cu' tuppu (granita with a warm brioche).

DETOUR

PRIMITIVO DI MANDURIA DOC
Puglia

Puglia has had a reputation for producing darkly colored, fruit-forward, richly alcoholic wines that enriched the lighter, less generous wines of northern Italy. Today, the region is working hard to distinguish itself from its bulk wine-focused past. A renewed interest in traditional grape varieties and unique terroirs makes this one of Italy's most dynamic wine regions.

Now trying to move away from bulk market past

THE PLACE

The Primitivo di Manduria DOC is in the Salento sub-peninsula (Italy's heel) in southern Puglia. The appellation's climate is very warm and dry, and the topography ranges from gently undulating to quite flat. Local soils contain high concentrations of iron oxide, giving a distinctive reddish hue to the landscape. Red grapes, especially Primitivo, thrive in these growing conditions.

CONSORZIO DI TUTELA
PRIMITIVO DI MANDURIA
DOC e DOCG

GRAPE VARIETIES

Primitivo (known as Zinfandel in the United States) is one of Puglia's flagship varieties. At the end of the 18th century, the grape was intentionally propagated by a priest near Gioia del Colle, who selected the vine because of its propensity to ripen very early. Eventually the grape acquired the name "Primitivo," most likely in reference to its tendency to ripen early. ("Primo" in Latin means "first.") Soon the variety could be found throughout Puglia.

PRODUCTION NOTES

min 85%

- The wines must contain a minimum of 85% Primitivo
- Riserva versions must be aged for at least two years, with a minimum of nine months in wood

WINE PROFILE

What makes this wine special: Primitivo produces fruit-forward, rich, ripe wines that are typically high in alcohol. The wines often possess a touch of sweetness which helps to balance the slightly rustic tannins.

Acidity: Medium

Tannin: Medium to high

Body: Full, with generous alcohol

Common descriptors: Cherry, raspberry, strawberry jam, macerated plums. In more traditionally styled wines, there are notes of tobacco and herbs.

Food pairings: Braised lamb, sausages, eggplant parmesan

NOTABLE PRODUCERS

Gianfranco Fino: This producer is responsible for showing the world Puglia's potential with his flagship Primitivo, "Es." Fino opted out of the DOC in 2015, to protest the relaxed regulations which allow producers to blend up to 15% of other varieties into their Primitivo di Manduria.

Luca Attanasio: Young producer employing traditional methods of viticulture and winemaking and completely dedicated to elevating the reputation of the Primitivo variety

PUGLIA'S SIGNATURE CUISINE

CHEESE

Burrata is made by filling a pouch of mozzarella with cream and bits of mozzarella curd. With its deliciously smooth, creamy texture, it is no wonder it earned the name "burrata," which means "buttery" in Italian.

BREAD

The Puglian city of Altamura has been famous for its palate-pleasing bread for centuries. First documented by the Latin poet, Horace, it is the only bread in all of Europe to be granted Protected Designation of Origin (PDO) status. The bread is made of 100% durum wheat grown in the province of Bari and ground into semolina. The loaves are generally shaped into a ball, which is scored with a knife. Once baked, the bread splits at the cuts and resembles a priest's hat (u cappidde de prèvete).

PASTA

Orecchiette is a centuries-old Apulian pasta tradition. Using the region's famous durum wheat, the pasta is characterized by its unique shape, which resembles "little ears," hence its name. The hollowed shape allows for plenty of sauce to be contained within each piece. The most traditional way to enjoy them is with cime di rapa (turnip tops), garlic and olive oil.

113

TAURASI DOCG
Campania

Taurasi lies in an area rich in viticultural tradition and was southern Italy's first DOCG. The wines of the appellation are made from Aglianico, the king of Campanian red grapes, widely considered to be one of Italy's three noble red grape varieties (alongside Nebbiolo and Sangiovese).

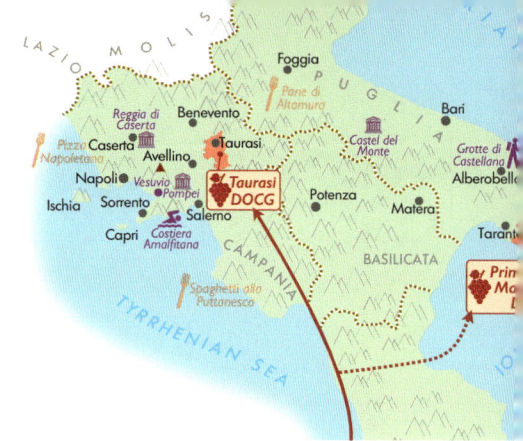

THE PLACE

The Taurasi DOCG lies within the Campanian wine district of Irpinia. The appellation's vineyards are planted at significant elevation in soils rich in volcanic materials, such as ash, tuff, lapilli and pumice. The warm days and cool nights of the area allow Aglianico to ripen slowly and build complex flavor compounds. This combination of climate, site and soil is ideal for Aglianico and the wines of the Taurasi DOCG are considered among the best expressions of the grape.

GRAPE VARIETIES

Aglianico (*Al-lee-AH-nee-koh*) is native to Campania and is thought to be one of Italy's oldest grape varieties. This thick-skinned grape requires a long growing season in order to fully ripen and produces particularly complex wines when grown on hillside vineyards with volcanic soils.

BEYOND TAURASI

Aglianico del Taburno DOCG
Campania is also home to the Aglianico del Taburno DOCG, another highly regarded appellation making wines from Aglianico. The DOCG lies in the Sannio wine district north of Irpinia; its wines show a bit more acidity than those of Taurasi.

Aglianico del Vulture
Aglianico is also an important grape for Basilicata and represents almost half of the land under vine in that region. For hundreds of years, it was widely accepted that the Aglianico growing in Basilicata was a completely different variety than the Aglianico in Campania. Although genetic testing has concluded that the two are actually the same grape, the National Registry of Grapes still lists them as separate and distinct.

The appellations of Aglianico del Vulture DOC and Aglianico del Vulture Superiore DOCG encompass fifteen villages that surround the ancient, extinct volcano of Vulture. The wines must be 100% Aglianico and tend to have a more intense red-fruit character than their Campanian counterparts.

Photo courtesy of Joaquin

WINE PROFILE

What makes this wine special: The variety is capable of producing long-lived wines that develop a great deal of complexity with bottle age. Its longevity, when combined with its powerful structural characteristics, has led to the nickname: "Barolo of the South."

Acidity: Bracing

Tannin: Powerful

Body: Full-bodied and high in alcohol

Common descriptors: In their youth, these are savory wines with aromas of sour cherry, violet, rose, red and blue berry fruit, licorice and cured meat. With bottle age, aromas of strawberry, dried fruit, leather and tobacco develop.

Food pairings: Rich, savory dishes, high in fat and intensity such as soppressata, braised lamb, osso buco, aged pecorino, and pasta with meat ragù

CONSORZIO TUTELA
VINI D'IRPINIA
Terra Verde

PRODUCTION NOTES

- The wines of the Taurasi DOCG must be a minimum 85% Aglianico; the majority, however, are 100% Aglianico
- The wines must be aged for a minimum of three years, with at least 12 months in wood
- Riserva versions must be aged for a minimum of four years, with at least eighteen months in wood

NOTABLE PRODUCERS

Mastroberardino: Iconic producer whose 1968 Taurasi Piano d'Angelo is widely regarded as one of Italy's greatest wines of all time

Contrade di Taurasi: Limited production of age-worthy 100% Aglianico wines that display power and purity

Cantine Guastaferro: Incredibly concentrated wines from vines that are 150–200 years old

Luigi Tecce: Traditionalist producer who shuns over-extraction and obvious oak in his wines

Perillo: Pre-phylloxera vines that yield powerful, structured wines with noticeable minerality

CAMPANIA'S SIGNATURE CUISINE

PASTA

Pasta Puttanesca
Stories abound about the origins of this savory dish—many of which are slightly off-color or ribald. The word "puttanesca" carries many connotations in the Italian language, from associations with prostitution to colloquial profanity. Regardless of the etymology, this dish is delicious. An umami-rich combination of tomatoes, olives, garlic, capers and anchovies is built into an aromatic and assertive sauce and ladled atop fresh pasta. The dish appeared in Naples in the mid-20th century and is closely associated with the city.

PIZZA

Pizza Napoletana
Napoli is known as the historic home of pizza. While flatbreads were a common Italian food for centuries, the tradition of topping it with tomatoes and cheese originated in Naples. The most traditional version is the "Margherita," a pizza that dazzles with its simple combination of five ingredients: locally made dough topped with tomatoes, mozzarella, basil leaves and a drizzle of olive oil. Italians take their Pizza Napoletana so seriously that they have formed the Associazione Verace Pizza Napoletana. This organization certifies restaurants that adhere to their official rules regarding authentic pizza production. There are only a few hundred restaurants in the world that are officially certified!

NOTES

APPENDIX

VINE TRAINING SYSTEMS

Italy employs a large variety of vine training systems. Viticultural practices have been shaped by the local environments. Though modern vine training and pruning systems are common across the entire country today, there are still certain areas that train vines using traditional systems.

The Etruscans introduced the first type of training systems. Vines were trained to grow high enough in trees so as to use branches as support. This system, known as alberata or vite maritata all'albero (married to the tree), remained widely practiced until the middle of the 20th century.

The Etruscan overhead vine training system is the ancestor of the more modern and traditional pergola and tendone systems still used today. Both of these systems allow the canopy to grow high above the ground and spread horizontally, with the grape clusters hanging underneath the horizontal trellis.

Pergola (IT)

Today the majority of Italian vineyards are trained using the standard vertical shoot positioning system whereby the fruiting canes are tied vertically to horizontal wires. Within this category are the Cordone Speronato and Guyot systems.

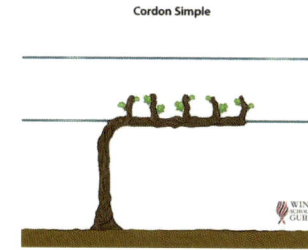

Cordon Simple

The Cordone Speronato is characterized by a permanent cordon of old wood (or two) trained horizontally.

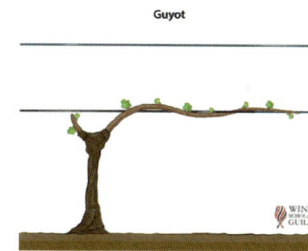

Guyot

Guyot is a common training system that has been used in Italy with all its variants for a long time. In this system, one or two new canes are kept every year and are trained horizontally. Regional variations such as Cappuccina, Capovolto and Archetto are also found.

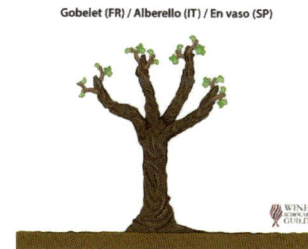

Gobelet (FR) / Alberello (IT) / En vaso (SP)

Alberello (or bush-trained vine) is a traditional low training system that is quite suitable for hot, arid climates. Vines are allowed to grow as free-standing, low bushes.

WINEMAKING

Making wine can be somewhat like a "choose your own adventure" story. At each stage in the process, there are a multitude of directions in which a winemaker can choose to go. There is no one correct way to make a wine! The following diagrams are intended to give the reader a basic understanding of the common processes used to craft red, white, rosé and sparkling wines. They also serve to hint at the myriad of decisions that today's winemakers have to make as they shepherd grapes from vine to wine.

SPARKLING WINE PRODUCTION

TRADITIONAL METHOD

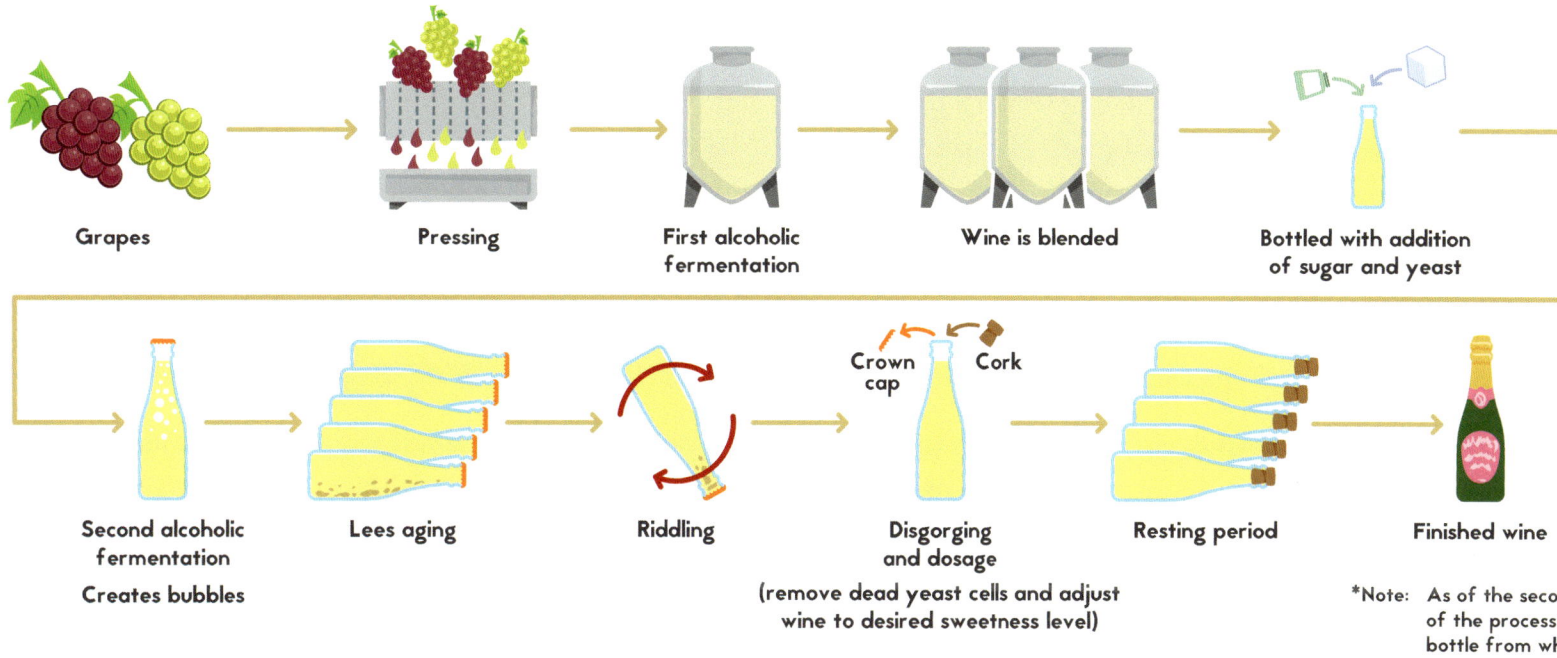

Grapes → Pressing → First alcoholic fermentation → Wine is blended → Bottled with addition of sugar and yeast

Second alcoholic fermentation

Creates bubbles

Lees aging

Riddling

Crown cap   Cork

Disgorging and dosage

(remove dead yeast cells and adjust wine to desired sweetness level)

Resting period

Finished wine

*Note: As of the second fermentation, the rest of the process takes place in the same bottle from which it is later served.

TANK METHOD

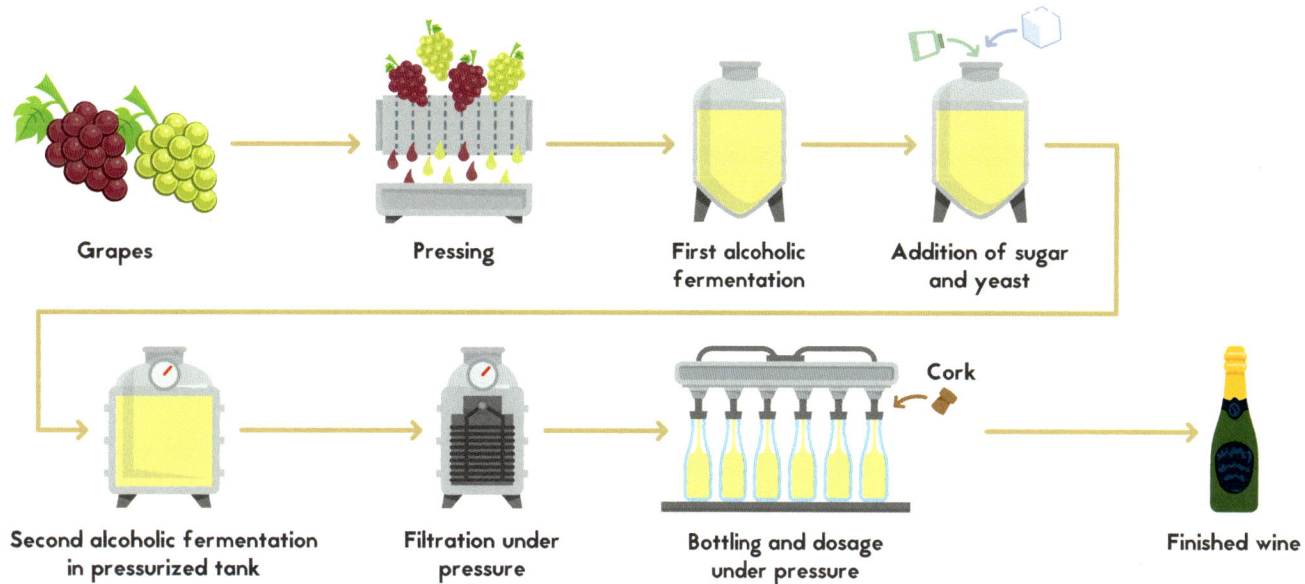

Grapes → Pressing → First alcoholic fermentation → Addition of sugar and yeast

Second alcoholic fermentation in pressurized tank → Filtration under pressure → Bottling and dosage under pressure   Cork → Finished wine

RED WINE PRODUCTION

WHITE WINE PRODUCTION

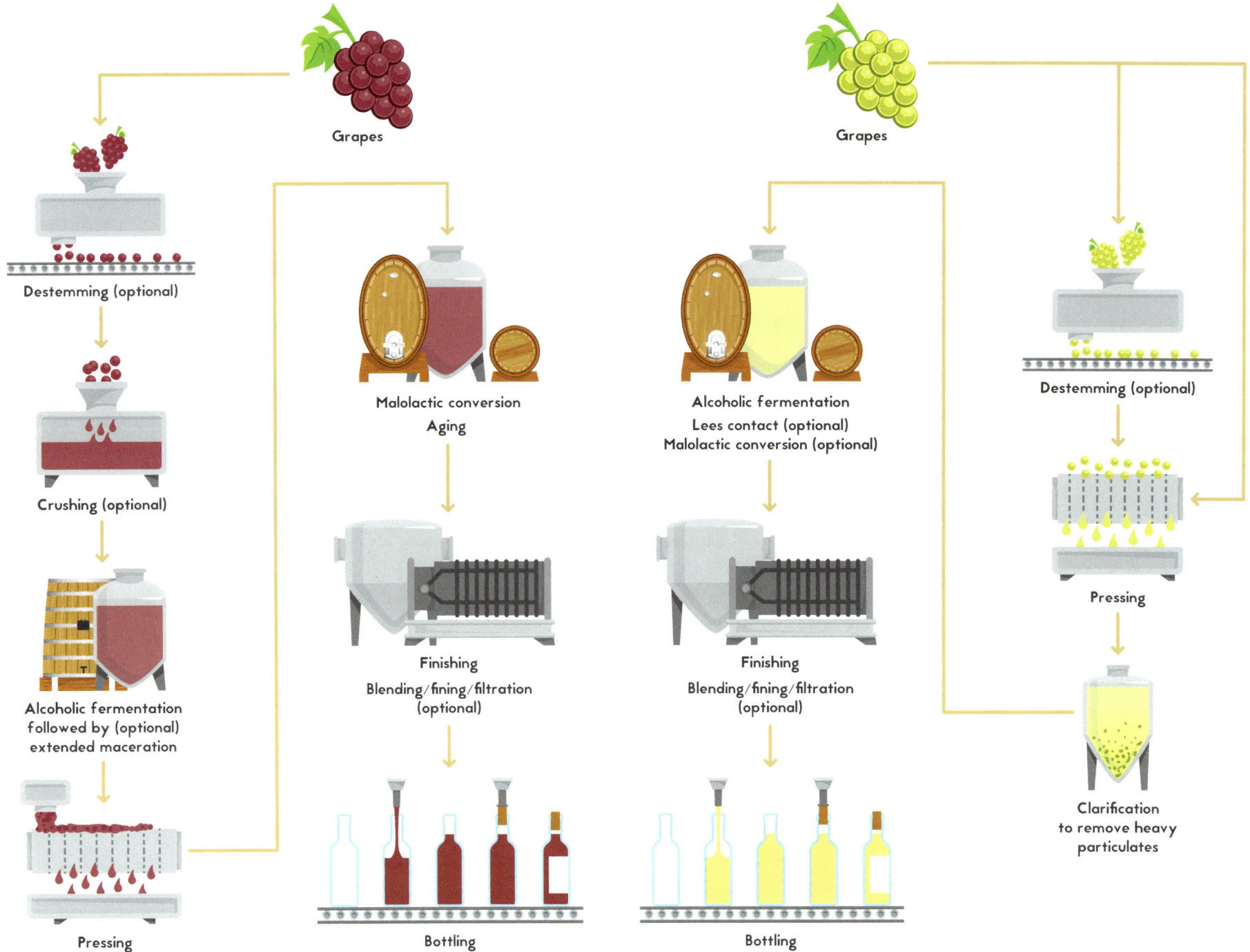

Grapes

Grapes

Destemming (optional)

Crushing (optional)

Alcoholic fermentation
followed by (optional)
extended maceration

Pressing

Malolactic conversion
Aging

Finishing
Blending/fining/filtration
(optional)

Bottling

Alcoholic fermentation
Lees contact (optional)
Malolactic conversion (optional)

Finishing
Blending/fining/filtration
(optional)

Bottling

Destemming (optional)

Pressing

Clarification
to remove heavy
particulates

ROSÉ WINE PRODUCTION

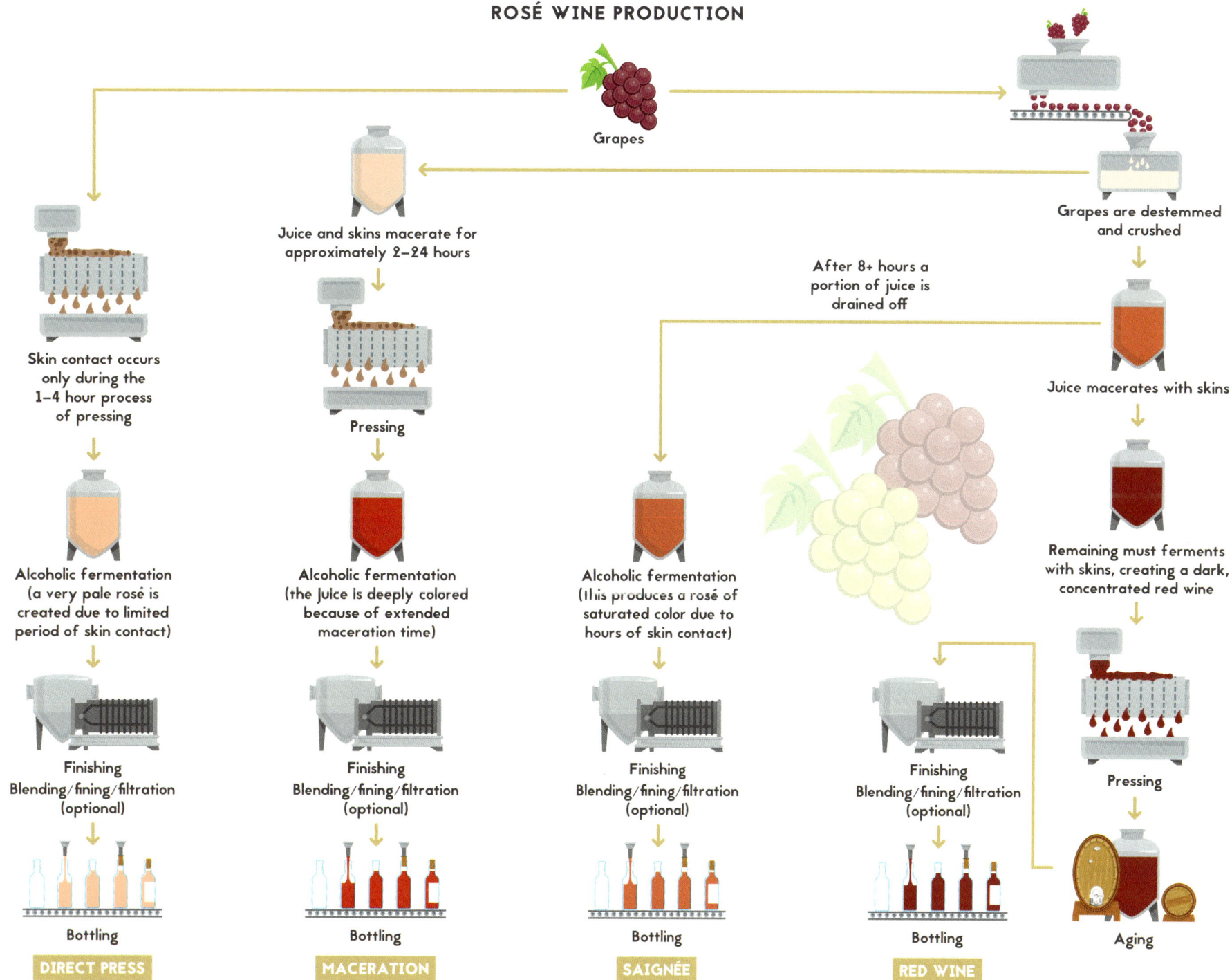

Grapes

Grapes are destemmed and crushed

Juice and skins macerate for approximately 2–24 hours

After 8+ hours a portion of juice is drained off

Juice macerates with skins

Skin contact occurs only during the 1–4 hour process of pressing

Pressing

Remaining must ferments with skins, creating a dark, concentrated red wine

Alcoholic fermentation (a very pale rosé is created due to limited period of skin contact)

Alcoholic fermentation (the juice is deeply colored because of extended maceration time)

Alcoholic fermentation (this produces a rosé of saturated color due to hours of skin contact)

Pressing

Finishing Blending/fining/filtration (optional)

Finishing Blending/fining/filtration (optional)

Finishing Blending/fining/filtration (optional)

Finishing Blending/fining/filtration (optional)

Aging

Bottling

DIRECT PRESS

Bottling

MACERATION

Bottling

SAIGNÉE

Bottling

RED WINE

123

NOTES

for all classic riserva if possible

get 1-2 and use Coravin

...LIST

try Rosso di Montalcino

WINE STYLES		CENTRAL ITALIAN WINES	SOUTHERN ITALIAN WINES
SPARKLING/ SWEET/ FORTIFIED WINES	Prosecco DOC/Prosecco Superiore DOCG *Trento DOC* * Franciacorta DOCG Saten Lambrusco DOCs Asti DOCG/Moscato d'Asti DOCG	*Vin Santo DOCs**	Passito di Pantelleria DOC Marsala DOC
WHITE WINES	Gavi DOCG Roero Arneis DOCG *Lugana DOC* * get Soave DOC *Alto Adige DOC** *Collio DOC & Friuli Colli Orientali DOC**	Vernaccia di San Gimignano DOCG get Orvieto DOC Verdicchio dei Castelli di Jesi DOC look for Riserva look for Orvieto Classico or Orvieto Classico Superiore also classico / classico superiore	Vermentino di Gallura DOCG def! Fiano di Avellino DOCG Greco di Tufo DOCG also Falanghina
RED WINES	Dolcetto d'Alba DOC Barbera d'Asti DOCG Barolo DOCG/Barbaresco DOCG Amarone della Valpolicella DOCG	Montepulciano d'Abruzzo DOC Chianti Classico DOCG Brunello di Montalcino DOCG riserva Bolgheri DOC Montefalco Sagrantino DOCG	do with cheese pairing Cannonau di Sardegna DOC Sicilia Nero d'Avola DOC - riserva Etna DOC riserva *Primitivo di Manduria DOC** Taurasi DOCG also riserva also try Cerasuolo di Vittoria DOCG

** Detour wine*

BECOME AN ITALIAN WINE SCHOLAR!

Upon successful completion of the IWS Prep exam you will be ideally positioned to pursue professional certification via the Italian Wine Scholar ™ (IWS) Certification Program.

The Italian Wine Scholar™ Certification Program is the industry's most respected certification course on the wines of Italy. The acclaimed curriculum is designed to provide committed students of wine with the most advanced and comprehensive specialization program on Italy bar none! It was created with the support of many of the Italian wine DOC/G consortia and has been endorsed by the Italian Trade Commission (ITA) in recognition of its exceptional level of depth, accuracy, detail, and academic rigor. Successful completion of the program confers an internationally recognized post-nominal (e.g. John Smith, IWS).

IWS is tailored toward individuals who are seeking to set themselves apart through specialization. Candidates are members of the wine trade, serious wine hobbyists and/or professionals making a career transition. The Italian Wine Scholar™ Certification Program is also a great resource and supplement for students moving toward advanced general wine study programs such as WSET Diploma (level 4), Master of Wine, the Court of Master Sommeliers' Advanced Sommelier (level 3 and up) or the Society of Wine Educators' Certified Wine Educator (CWE).

The IWS curriculum guides students through an in-depth exploration of Italy's 20 distinct wine regions, within the context of:

- History, culture and traditions
- Location and climate
- Topography, geography and soils
- Grape varieties and their organoleptic profiles
- Viticulture and winemaking
- Wine industry trends and economics
- All DOCs and DOCGs with their regulations, wine styles and specificities

While the sheer volume of information necessary to master Italian wine is extraordinary, the course material is broken down into two manageable chunks. There are two separate study manuals (Unit 1: Northern Italy and Unit 2: Central & Southern Italy) and two separate exams. The Italian Wine Scholar™ credential/pin is based on the final composite score of both exams. The units can be taken in any order.

The exam for each unit is composed of 100 multiple-choice questions. Candidates need a composite score of 75% or higher to receive the credential, i.e. the scores from both exams are added together and averaged for final assessment.

Students with a composite score of 85-90 pass with HONORS. Candidates scoring 91-100 pass with HIGHEST HONORS.

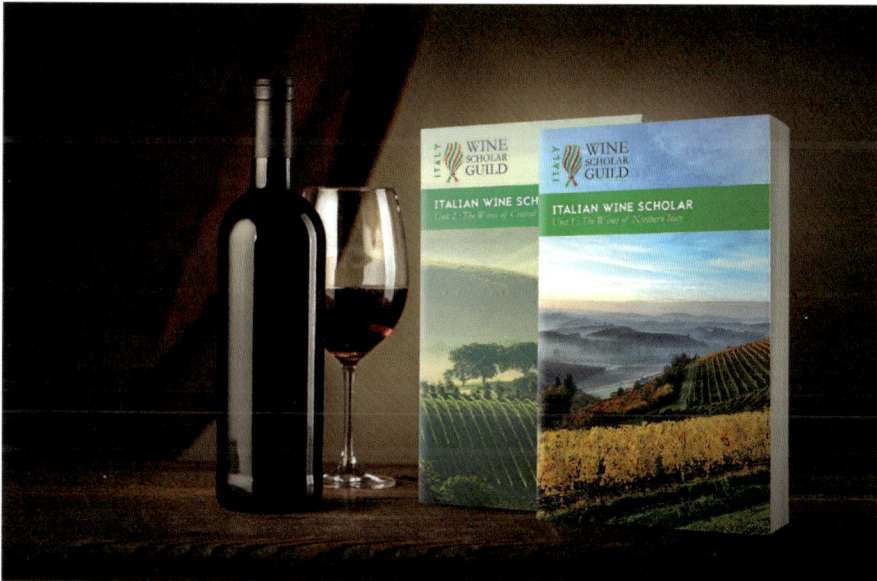

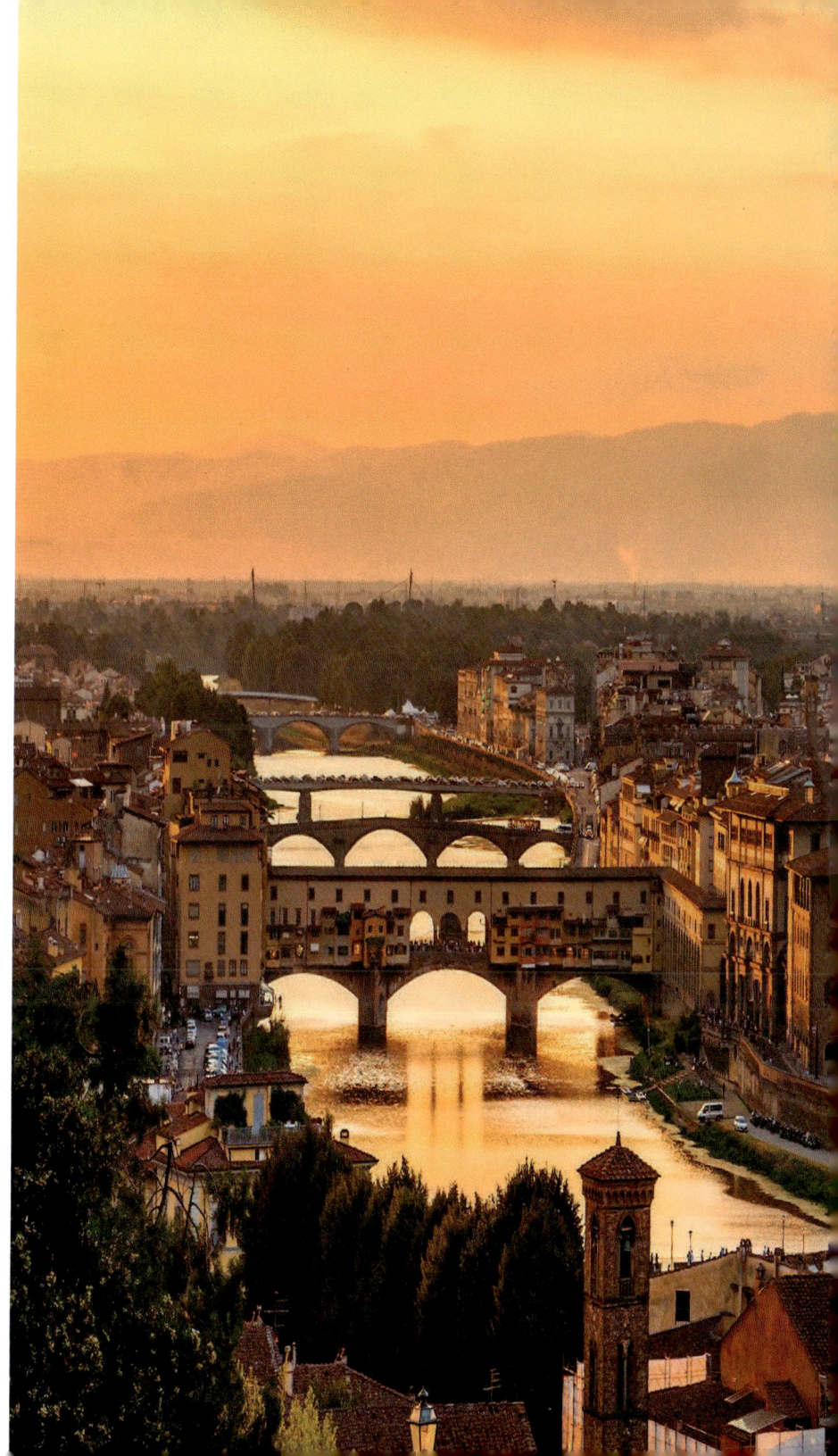

WINE STUDY IS A JOURNEY...KEEP YOURS IN MOTION!

For more information please visit us at
www.winescholarguild.org/iws

Edition 1.1